Beyond Hearing Aids: Targeting the Cognitive Roots of Speech in Noise Difficulties

Saqlin

Table of Contents

Chapter 1: Introduction 10

Chapter 2: Literature Review 15

    Early Studies of Auditory Selective Attention 15

    Executive Functions, Effort, Load and Distractibility 19

        WMC and Distractibility 20

        Effort and Distractibility 27

        Load and Distractibility 31

    Motivation for the Current Work 38

        Specific Aims and Hypotheses 42

Chapter 3: Methods 44

    General Methods 44

        Participants 44

        WMC 45

        General Procedures 46

        Stimuli 47

        Distractor Paradigm 48

    Experiment I – Procedure 49

    Experiment II – Procedure 50

Chapter 4: Results 53

    Manipulation of Load 53

    Background Noise Conditions 57

    Distractibility 59

    WMC Groups 61

Chapter 5: Discussion 71

    Experiment I 72

Experiment II 81

General Discussion 89

Chapter 6: Conclusion... 98

References... 101

Chapter 1: Introduction

The chief complaint among listeners seeking clinical methods of audiologic rehabilitation is difficulty separating speech from background noise. While reports are uniform, experiences are variable and subject to interactions between both peripheral (i.e., bottom-up) and central (i.e., top-down) factors. Recent cognitive hearing science studies have explored how central factors such as working memory capacities (WMCs) and control mechanisms (i.e., attentional and inhibitory) contribute to the *repair* of breakdowns imposed by peripheral factors such as hearing impairments (Arlinger, Lunner, Lyxell & Pichora-Fuller, 2009; Humes, Kidd & Lentz, 2013; Pichora-Fuller et al., 2016; Rönnberg et al., 2013; Strauss & Francis, 2017); however, greater insights to *how* and *why* breakdowns occur are necessary to achieve a holistic understanding of these listener complaints. Perhaps revisiting studies of selective attention through the lens of auditory distraction would begin to fill-in these gaps.

Broadly stated, selective attention refers to the mechanism responsible for the separation of intended inputs from irrelevant inputs. With this definition in mind, listener difficulties amidst competing noise can then be attributed to failures of selective attention (Dai, Best & Shinn-Cunningham, 2018; Murphy, Groeger & Greene, 2016; Oberfeld & Klöckner-Nowotny, 2016; Shinn-Cunningham & Best, 2008). This phenomenon of unsuccessful separation allows for intrusions from irrelevant inputs, or instances of auditory distraction.

Magnitudes of interference, or measures of distractibility, have been shown to correlate to the level of processing irrelevant inputs reach (Craik, 2014; Driver, 2001;

Gomes, Barrett, Duff, Barnhardt & Ritter, 2008; Hughes, Hurlstone, Marsh, Vachon & Jones, 2013), to which selective attention has been described as a "gatekeeper" (Awh, Vogel & Oh, 2006; Sörqvist, Stenfelt and Rönnberg, 2012). Within the "gatekeeper" framework, selective attention either prohibits distractions from entering focal attention or grants their entry. Models outlining the functionality of selective attention have been proposed, yet remain debated and unresolved within the auditory domain (Murphy, Groeger & Greene, 2016; Spence & Santangelo, 2010; Shinn-Cunningham & Best, 2008).

Three of the more commonly referenced models of selective attention are: early selection, late selection, and attenuation (Driver, 2001; Lavie & Tsal, 1994). The *early* camp argues that features of relevant inputs are identified at lower peripheral levels of processing, and inputs that do not share the same qualities (i.e., distractors) are excluded from reaching higher or more central levels (Broadbent, 1958). In contrast, the *late* camp suggests that all inputs are perceived (i.e., both relevant and distracting) and must be mediated later with the assistance of central mechanisms such as inhibitory control (Deutsch & Deutsch, 1963). Attenuation views propose that early selection occurs until a salient distractor is introduced, which then must be segregated later (Treisman, 1964). As attention research matured, it became clear that one model could not account for all scenarios of distraction alone (Francis, 2010; Lavie, 2010; Lavie, Beck & Konstantinou, 2014; Murphy, Groeger & Greene, 2016). Instead of pitting these models against each other, the dialogue switched following the introduction of load theories towards understanding how demands of tasks or objectives influence methods of selection as well as experienced distractibility (Driver, 2001; Forster & Lavie, 2009; Lavie, 2010; Lavie, Hirst, de Fockert & Viding, 2004; Lavie & Tsal, 1994; Murphy, Groeger & Greene, 2016; Spence & Santangelo, 2010).

Per load theory accounts, tasks may have one of two forms of load (*cognitive* or *perceptual*) and attention is defined as a resource dependent mechanism (Lavie & Tsal, 1994). *Cognitively* loaded tasks are assumed to induce *mental* difficulty, meaning mechanisms such as working memory must be recruited. As cognitive demands increase and tax central functions, individuals become more susceptible to distraction as fewer resources – if any – are left to mediate intrusions. On the other hand, *perceptual* load is defined by the complexity of bottom-up features included or needed to identify target-objects within tasks. The constraint of attention towards target-object identification under high perceptual load has an opposite effect on distractibility than cognitive load, in that the lack of residual focal attention prevents recognition of distracting inputs (Francis, 2010; Lavie, 2010; Lavie, Beck & Konstantinou, 2014; Lavie, Hirst, de Fockert & Viding, 2004; Lavie & Tsal, 1994; Simon, Tusch, Holcomb & Daffner, 2016). A caveat of the load theory is that effects of both forms of load have yet to be demonstrated and agreed upon in the auditory domain (Murphy, Fraenkel & Dalton, 2013; Murphy, Groeger & Greene, 2016; Murphy, Spence & Dalton, 2017; Tan et al., 2015); most supporting evidence has been derived from visual and/or cross-modal paradigms.

An exception to the latter statement is that cognitive load has been discussed within listening effort literature. Effortful listening results from complex auditory environments and/or in the presence of degraded speech (Peelle, 2018; Pichora-Fuller et al, 2016). General findings support assumptions of the cognitive load theory, as recruitment of cognitive *effort* to restore breakdowns in speech tends to result in poorer listening performance measured both behaviorally and via self-report (Lemke & Besser, 2016; Peelle, 2018; Pichora-Fuller et al, 2016; Strauss & Francis, 2017). While instances of effortful listening are irrefutable, contributions from solitary auditory *perceptual* demands remain largely unknown.

To reiterate, increases in *perceptual* load have been shown to reduce measures of distraction across visual and cross-modal paradigms, and definitive support specific to the auditory domain (i.e., auditory task – auditory distractors) has yet to be provided (Gomes, Barrett, Duff, Barnhardt & Ritter, 2008; Fairnie, Moore & Remington, 2016; Francis, 2010; Murphy, Fraenkel & Dalton, 2013; Murphy, Spence & Dalton, 2017; Spence & Santangelo, 2010). Nonetheless, results following auditory perceptual load manipulations within cross-modal designs as well as awareness reports *may* indicate that predictions can transfer to the auditory domain (Macdonald & Lavie, 2011; Murphy & Greene, 2017). For example, Murphy and Greene (2017) assessed driver awareness of road obstacles in the presence of auditory perceptual load where load was manipulated by changing the objective assessment of a radio stream. Under high load participants were asked to identify when a major news event was announced in the radio stream, while under low load they only listened for changes in talkers. When auditory perceptual load was high, participants exhibited reduced awareness or perception of visual obstacles introduced into the roadway (Murphy & Greene, 2017). While findings reflected successful manipulation of auditory perceptual load and resulted in patterns similar to visual load, effects were still cross-modal and general assumptions of auditory selective attention could not be made.

Given the importance of auditory selective attention for listening in adverse conditions, it is crucial to reinvestigate this mechanism in isolation. Perceptual load studies offer a solution to this, as their primary objective is to limit the engagement of higher order functions such as working memory, attentional and inhibitory control. Following predictions from visual designs, irrelevant sounds access to focal attention should be limited in the presence of high demands; however, these scenarios are where listeners experience the most difficulty. The present study has been designed to evaluate how bottom-up auditory perceptual demands directly influence interactions between irrelevant distractor sounds and focal attention in the absence of cognitive load across two listening tasks. Additionally, environmental factors such as the type of background noise and individual characteristics such as WMCs are known to contribute to performance differences under *cognitive* load; therefore, these have been included within the present study to gain an exploratory look at how they interact with lower-level *perceptual* demands.

Chapter 2: Literature Review

Auditory distractions are commonly encountered in our daily environments, whether they result from a friend calling your name, the rustling of a student late to a lecture, or the receipt of a phone call; however, distractions become undesirable when they impair abilities to achieve goals and perform tasks. Concerning clinical audiology and cognitive hearing science, factors such as the presence of background noise (DiGiovanni, Riffle, Lynch & Nagaraj, 2017; Herweg & Bunzeck, 2015; Trimmel, Schätzer & Trimmel, 2014), individual differences in executive processes (Rönnberg et al., 2013; Sörqvist & Rönnberg, 2014), and type of task load (Lavie, 2010; Lavie & Tsal, 1994) have all been identified as contributors to variations in distractibility or performance on listening based tasks. Although these factors have been recognized, compounding effects along with their potential influences over the efficiency of selective attention are rarely discussed in unison. Here we will review early studies of selective attention, emerging work regarding individual differences and task demands, as well as unresolved questions posed by the field.

Early Studies of Auditory Selective Attention

The earliest and most commonly referenced studies of auditory selective attention and listening in background noise stem from Cherry's (1953) historic discussion of the "cocktail party effect" (Bronkhorst, 2015; Driver, 2001). In short, the "cocktail party effect" suggests that listeners can identify features of sounds that are important and/or interesting while simultaneously filtering out sounds unrelated to their main goals. In other words, if a listener intends to converse with a talker amongst background conversations, features of that talker's voice can be separated via selective attention from the additional noise (Bronkhorst, 2015; Cherry, 1953). Further support for this phenomenon was provided through dichotic listening tasks, where two separate streams of information (i.e., to-be-attended-to and to-be-ignored) were presented to each ear simultaneously. Cherry (1953) demonstrated the ability to ignore the irrelevant steam as participants were not able to report any context from the unattended messages other than the occasional change in talkers or addition of non-speech sounds (i.e., tones) (Bronkhorst, 2015; Driver, 2001; Spence & Santangelo, 2010).

This ability to separate, or filter, background noise from intended speech was addressed by Broadbent (1958) through the proposal of the filter theory of attention. Per the filter theory, acoustic environments are surveyed for features that may become important for a listener. Once target information is identified, it is passed through a metaphorical attentional filtering system. The *filter* is not limitless, so additional sounds not entered into the filter (i.e., distractors) decay as time goes on and never reach higher levels of processing; in this case *early* selection of distractors occur (Broadbent, 1958; Driver, 2001; Spence & Santangelo, 2010). Within the framework of this theory, only sounds that are relevant to a listener are brought into focal attention and extraneous background noise is easily separated out at a lower peripheral level. While this account fits Cherry's (1953) conclusions well, it later became clear that the filter theory of attention could not explain selective attention alone (Driver, 2001).

Moray (1959) then expanded upon initial "cocktail party" findings to identify when listeners' attention may be pulled off their target speaker. Like Cherry (1953),

Moray (1959) found that listeners could only attend to a portion of the competing background noise; however, when a listeners' name was introduced into the background it became salient enough to pull attentional resources away from the target speaker. This result suggested that Broadbent's filter theory was not an airtight account of selective attention (Driver, 2001; Murphy, Groeger & Greene, 2016; Spence & Santangelo, 2010). Instead, Moray's findings appear to align with the attenuation theory (Treisman, 1964); which states that all inputs are initially perceived in some form, yet the perception of irrelevant sounds can be decreased unless a dramatic change (i.e., alarm or increase in volume) is noted. Framing Moray's findings with the attenuation theory suggests that the calling of one's name is a sufficient change that constitutes a redirection of attention.

The latter argument also fits within Deustch and Deustch's (1963) *late* selection account of selective attention, which again posits that all inputs are initially perceived yet become separated later with the assistance of executive functions. Even though Deustch and Deustch's view proposed different timing of distractor mediation, the way target inputs were eventually selected was akin to the latter two camps; target information is identified via relevance to a current goal (Spence & Santangelo, 2010). This late selection concept also calls for discussions of control mechanisms such as inhibition, as inhibitory control refers to the mechanism that allows for the reduction of negative effects from irrelevant inputs or representations during encoding or retrieval of relevant information (Anderson, 2001; Hasher, Lustig & Zacks, 2007).

Introducing inhibition to the discussion of these filter and attenuation theories pulls in additional executive functions that may play a role. Conway, Cowan, and

Bunting (2001) later explored the "cocktail party effect" in terms of working memory. Cowan's initial perspective of working memory proposed three major components and a limited capacity of roughly three to five items that can be entered or stored at a time: 1) the focus of attention, 2) activated subset of long-term memory, and 3) connections to long-term memory (Cowan, 1999; 2010). With the acknowledgement of working memory capacity (WMC) limitations, "cocktail party effects" were investigated with respect to individual differences in capacity. Findings supported the notion that individuals with lower WMCs, or greater limitations, were more likely to notice their names amongst background noise than high-capacity counterparts. This was later interpreted as individuals with greater limitations (i.e., lower WMCs) have poorer attentional control and constraint over focal attention (Cowan et al, 2005; Kane, Bleckley, Conway & Engle, 2001; Macken, Phelps & Jones, 2009).

Following new conclusions and perspectives of the "cocktail party effect", researchers became interested in additional ramifications of differences in executive functioning (Arlinger, Lunner, Lyxell & Pichora-Fuller, 2009; Peelle, 2018; Rönnberg et al., 2013). From this line of work, implications for real-world listening and perceived mental effort studies emerged (Francis, Bent, Schumaker, Love & Silbert, 2021; Pichora-Fuller et al., 2016; Sörqvist & Rönnberg, 2014). Again, the major executive functions explored within this realm have been working memory, inhibition, and attentional control.

Executive Functions, Effort, Load and Distractibility

The processes of working memory, inhibition, and attentional control are frequently referenced in cognitive literature; however, definitions of their main functions can get lost in translation (Cowan, 2017; Oberauer, 2019). The following reflect my interpretations of these processes for the purpose of this review: 1) working memory is responsible for concurrent storage, manipulation, and maintenance of information or goals over brief periods in time (Cowan, 2017), 2) inhibitory control provides listeners with the ability to reduce negative effects of task-irrelevant stimuli (Anderson, 2001; Hasher, Lustig & Zacks, 2007), and 3) attentional control broadly reflects the ability to allocate attentional resources and serves as an umbrella term for divided, sustained, and selective attention (Engle, 2018; Hasher, Lustig & Zacks, 2007). Successful completion of demanding tasks requires coordination between these systems.

Consider the following example: a student is listening to a lecture presented in a hall where other students are typing, talking, and moving about. The student must engage attentional control to direct focal attention towards the lecturer and away from distracting sounds. Should these unwanted sounds gain access to focal attention, the student must then employ inhibition to reduce their effects. While attentional control is directing focus and inhibition is suppressing interference, working memory is diligently storing main ideas from the lecture to form holistic thoughts and transfer these to long-term memory. Additionally, working memory assists with the repair of breakdowns of intended messages should attentional or inhibitory control allow distractors to interfere. Even though the given example highlights a single student, these processes are working in the background for everyone attending the lecture; what one may find distracting may not be for another. This variability in distraction has often been tied to differences in WMC limitations (Fougine, 2008; Kane & Engle, 2003; Rönnberg et al., 2013; Sörqvist, 2010).

WMC and Distractibility

Like the aforementioned executive functions, WMC has also been discussed with more than one definition (Burgoyne & Engle, 2020): 1) the finite number of items that can be entered or stored in working memory (Cowan, 2010; Miller, 1956), and 2) WMC reflects the ability to allocate or coordinate attentional resources towards task completion (Cowan, 2017; Oberauer, 2019; Unsworth & Engle, 2007). The latter definition is supported by the controlled attention view of WMC (Kane, Bleckley, Conway & Engle, 2001; Engle, 2018) and will be used for further discussions here. With this interpretation in mind, individuals with high-WMCs have expressed superior attentional control abilities compared to those with low (Kane, Bleckley, Conway & Engle, 2001; Shipstead, Harrison & Engle, 2015; Sörqvist, Nöstl & Halin, 2012); this concept was also introduced in the previous section following revised evidence for the "cocktail party effect" (Conway, Cowan & Bunting, 2001; Macken, Phelps & Jones, 2009). Regardless of the definition adopted by researchers, complex span tasks are routinely administered as measures of WMC (Simon, Tusch, Holcomb & Daffner, 2016; Sörqvist, 2010; Wilhelm, Hildebrandt & Oberauer, 2013).

Common forms of complex span measures are reading (RSPAN), listening (LSPAN), operation (OSPAN) and backward digit spans (Daneman & Carpenter, 1980; Mathy, Chekaf & Cowan, 2018). A key feature among these tasks that make them appropriate for computing WMCs is that they require both a processing (i.e., encoding or judgement) and storage component for later recall (DiGiovanni, Riffle, Weaver & Lynch, 2017; Rönnberg et al, 2013; Shipstead, Harrison & Engle, 2015; Sörqvist, 2010; Wilhelm, Hildebrandt & Oberauer, 2013). Variations of these span tasks have been applied to attempt to predict measures of listeners' distractibility across auditory distractor paradigms.

External distractors may influence performance via interference-by-process, attentional capture, response competition, or a combination of these events. Instances of interference-by-process are observed in the presence of changing-state sounds, where distractors are presented in a similar manner to target stimuli. Negative changing-state effects are predominantly noticed during serial recall, and WMCs generally do not predict performance as the mechanism contributing to detriments in serial recall has been argued to be *task* dependent (i.e., the need to process inputs in a specific order that is impaired by distractors presented in a similar order) instead of directly attention based (Hughes, 2014; Hughes, Hurlstone, Marsh, Vachon & Jones, 2013; Macken, Phelps & Jones, 2009; Marois, Marsh & Vachon; Sörqvist, 2010, 2010; Sörqvist, Marsh & Nöstl, 2013). Rather than solely influencing performance on a specific sub-set of tasks (i.e., serial recall), consequences of attentional capture and response competition have been argued to be more universal.

Attentional capture is elicited by the introduction of an *unexpected* or *deviant* sound, as seen within oddball paradigms (i.e., a "standard" sound is presented 80-90% of trials, with a "deviant/oddball" heard the remaining 10-20%). Once attention is "captured" listeners must reorient or redirect focal attention back to intended inputs; this time spent away from intended inputs results in missed information and serves as a detriment to performance (Hughes, 2014; Hughes, Hurlstone, Marsh, Vachon & Jones, 2013; Macken, Phelps & Jones, 2009; Marois, Marsh & Vachon, 2019; Sörqvist, 2010, 2010; Sörqvist, Marsh & Nöstl, 2013). In the case of response competition, distractibility depends on the relevance between distracting and target inputs (i.e., distractors similar to intended inputs are more detrimental) and is frequently observed via Stroop (Color-Word identification tasks; Kane & Engle, 2003; Meier & Kane, 2013; Stroop, 1935) or Erikson flanker paradigms (distractors are presented outside of a central task; Erikson & Erikson, 1974; Lavie, 2010; Ulrich, Prislan & Miller, 2021). Along with time of reorientation following "capture", relevant distractors entered into focal attention can now interact with judgements of intended inputs; hence the "response competition" title (Erikson & Erikson, 1974; Kane & Engle, 2003; Lavie & Tsal, 1994; Meier & Kane, 2013; Tan et al., 2015). In addition to affecting performance beyond serial recall, outcomes of attentional capture and response competition appear to be mediated by differences in WMCs which has been demonstrated across multiple behavioral and objective (i.e., physiologic) investigations.

For example, Sörqvist (2010) divided participants into high- and low-WMC groups via OSPAN scores and tasked them with the recall of printed digits. This recall task was performed in the presence of three auditory distractor conditions: steady-state (all distractor sounds were the same), changing-state (strings of spoken consonants or tones with various frequencies), and deviant stimuli (an unexpected consonant or tone was entered into otherwise steady-state streams). WMC group differences were identified during the deviant (i.e., oddball) condition, yet - unsurprisingly - did not emerge during the changing-state condition. In the presence of deviant distractors, those with high-WMC were able to negate effects of attentional capture while the low-WMC group was unable to (Sörqvist, 2010). This superior mediation of deviant sounds by high-WMC participants was replicated in an additional study done by Sörqvist, Nöstl and Halin (2012).

Within the follow-up investigation, participants were tasked with visual arrow identification while listening to auditory stimuli; an auditory oddball (i.e., deviant) distractor paradigm was employed. Arrows were presented across six blocks and participants were to decide if the target-arrow was oriented towards the right or left. For the low-WMC group, reaction times (RTs) were significantly slower in the presence of deviant sounds; this significant increase in RT persisted across all six blocks. For the high-WMC group, RT differences disappeared across blocks indicating attenuation of the orienting response. These results provided further behavioral support for superior control over the orienting response for individuals with higher WMCs (Sörqvist, Nöstl & Halin, 2012). It is important to note here that deviant distractors across the two examples above shared no meaningful overlap with target stimuli (i.e., Example 1: target = printed digit, deviant = spoken consonant/tone (Sörqvist, 2010); Example 2 (Sörqvist, Nöstl & Halin, 2012): target = visual arrow, deviant = tone); therefore, consequences following attentional capture were only temporal in nature reflected by increases in time pulled away from the perception of target stimuli. During real-world endeavors, however, it is likely that individuals will encounter distractions that are meaningful and may share overlap with desired inputs (Forster & Lavie, 2009). Should attention be directed towards *meaningful/related* distractors, we see response competition (Erikson & Erikson, 1974; Heitz & Engle, 2007; Kane & Engle, 2003; Meier & Kane, 2013; Oberfeld & Klöckner-Nowotny, 2016).

Response competition relies on distractor relationships to targets, where distractors share one of two or three relationships to a given target: congruent (compatible/same as), neutral (no relationship), or incongruent (incompatible/response competing); however, the definition of these relationships varies depending on the paradigm put in place by researchers. For example, Dalton, Santangelo and Spence (2009) defined congruency by spatial location, while Stroop (1935), Kane and Engle (2003) determined congruency by stimulus features paired with task requirements (i.e., congruent/compatible – "BLUE" is written/presented in *BLUE* text, incongruent/incompatible – "BLUE" is written/presented in *RED* text; Meier & Kane, 2013), and Li and Lou (2019) as well as Lavie (2010) deemed congruency via overlap with a given target (i.e., target letters are "x" or "z", distractors are either compatible (X – x, Z – z), incompatible (X-z, Z-x), or neutral (N-x or z)). Regardless of the competition paradigm employed, participants perform worse reflected through longer RTs and poorer accuracy when an incongruent/incompatible distractor is present. Additionally, when WMCs have been included as a factor within experimental designs, individuals with higher capacities show superior abilities to mitigate effects of response competing or incongruent distractors than those with low-WMC consistent with findings from attentional capture work (Dalton, Santangelo & Spence, 2009; Kane & Engle, 2003; Meier & Kane, 2013). This superior mediation of response competing information also sheds light on what listeners may do and/or how WMCs relate to the ability to "handle" information that reaches Cowan's idea of an activated subset of long-term memory; however, the discussed distraction studies only look at how the ability to identify a target is impaired, leaving less known about what specific levels of processing they were eventually mediated at. To gain better insights to the latter phenomenon, objective or perceptual measures have become particularly important.

Objective measures provide data regarding the fate of internal processing and levels that distractors reach through event-related potential (ERP) assessments such as the auditory brainstem response (ABR), mismatch negativity (MMN), and P300 (Gomes, Barrett, Duff, Barnhardt & Ritter, 2008). For example, Sörqvist, Stenfelt and Rönnberg (2012) presented participants with a working memory n-back task and collected ABR recordings to distracting tones. Participants of this study were also divided into high- and low-WMC groups via OPSAN scores and performance on a size comparison task. The high-WMC group expressed significantly larger reductions in wave-V amplitude compared to the low-WMC group (Sörqvist, Stenfelt & Rönnberg, 2012). The benefit of ABR recordings is that they shed light on how auditory distractors are perceived at peripheral or *early* levels of processing (e.g., wave-V originates from the lateral lemniscus tract of the brainstem; Lewis, Kopun, Neely, Schmid & Gorga, 2015). What is interesting from Sörqvist, Stenfelt and Rönnberg (2012) results is that differences in top-down functions (i.e., WMCs) extended downwards and influenced bottom-up perception of these auditory distractor sounds; MMN and P300s then become useful to measure higher-order perceptual events (Rönnberg et al, 2013).

P300s gauge active attentional processes, while MMN recordings do not require active attention and measure the auditory system's innate ability to detect changes in sounds during oddball (i.e., deviant) paradigms (Saliasi, Geerlings, Lorist & Maurits, 2013; Sur & Sinha, 2009; Näätänen, Sussman, Salisbury & Shafer, 2014). Many of these studies compare responses from older and younger adults; however, research has demonstrated that older adults have poorer attentional control and lower WMCs than younger adults (Atcherson, Nagaraj, Kennett & Levisse, 2015; Rönnberg et al., 2013). Overall, both P300 and MMN responses to deviant stimuli tend to be significantly reduced when collected from older groups of participants (Dong, Reder, Yao, Liu & Chen, 2015; Erikson, Ruffle & Gold, 2016; Saliasi, Geerlings, Lorist & Maurits, 2013); corroborating arguments of poorer distractor mediation through increased deviation effects for low-WMC listeners. With these findings from objective measure assessments in mind, it is likely that higher capacity listeners can engage selective attention *early* (Broadbent, 1958; Bronkhorst, 2015) while lower capacity listeners with poorer attentional control cannot enact these processes *early* and instead do so *later*. Additionally, should distractors undergo *late* selection for high-WMC listeners, this group of listeners react more effectively (Deustch & Deustch, 1963; Kane & Engle, 2003; Meier & Kane, 2013; Sörqvist, Marsh & Nöstl, 2013; Treisman, 1964)

General conclusions from theses behavioral and objective findings, apart from changing-state effects specific to serial recall, support the controlled attention view of

WMC and argue that low-WMC listeners are more susceptible to failures of selective attention via increased experimental RTs and reduced integrity of neural responses. Additionally, high-WMC listeners appear to be better at mediating representations in focal attention for later performance (i.e., deployment of inhibitory control) – which may become necessary in the face of response-competing distractors. What is lacking from the narrative at this point, are real-world consequences of distraction and contributions from individual differences in WMCs as well as control mechanisms. Mentioned in the introduction, real-world outcomes are frequently assessed through reports and measures of listening effort (Francis, Bent, Schumaker, Love, & Silbert, 2021; Guijo, Horiuti, Nardez & Cardoso, 2018; Rudner, Lunner, Behrens, Thorén & Rönnberg, 2012; Strauss & Francis, 2017).

Effort and Distractibility

To reiterate, listening effort refers to the amount of cognitive energy expended to complete a complex listening task (Guijo, Horiuti, Nardez & Cardoso, 2018; Pichora-Fuller et al, 2016). While the validity of listening effort measures and definitions are still being ironed out (Francis & Love, 2019; Pichora-Fuller et al., 2016), the purpose of including this "Effort and Distractibility" section was to provide a holistic picture of experiences as well as mechanisms engaged during *complex listening*. Within this line of work, *distractibility* is not measured directly through instances of interference by singular sounds (i.e., deviants); instead, effects of background noise, peripheral hearing sensitivity, and signal processing (i.e., amplification devices) are discussed. Regardless of the perceptual cause (e.g., deviance, background noise, or distortion), outcomes of failures of selective attention remain largely the same in that some portion of intended auditory inputs, or speech in the case of listening effort research, are missed (Shinn-Cunningham & Best, 2008; Strauss & Francis, 2017). Consistent with trends of distractibility discussed thus far, differences in executive functions correlate to both exerted effort and negative effects from interference.

To broadly explain the repair process following breakdowns of intended speech inputs in the context of working memory, Rönnberg et al (2013) proposed the Ease of Language Understanding (ELU) Model. Per the simplified ELU model, speech inputs are processed through a rapid, automatic multimodal binding of phonology (RAMBPHO) system. Messages that "match" long-term memory constructs flow quickly through this system, while inputs resulting in "mismatches" require additional reconciliation from an explicit processing loop. The ELU and Cowan's working memory model do not describe the same mechanism, yet Cowan's embedded processes components play a major role for ELU predictions. For example, inputs are brought into awareness via Cowan's idea of the focus of attention; items in awareness can then be identified as relevant or irrelevant and offloaded to their respective storage areas. Flow of information from the focus of attention to long-term memory is therefore more successful when there is a "match" in Rönnberg's terms (Rönnberg et al, 2013; Rönnberg, Holmer & Rudner, 2019). The amount of signal degradation perceived relates to the number of resources, or amount of effort, needed for the explicit processing loop to resolve these breakdowns.

Since the ELU model describes working memory's role during speech understanding, contributions from individual differences in WMC are naturally included within listening effort discussions. Generally, listeners (with and without hearing-impairments) with higher WMCs have shown better performance on complex listening tasks compared to those with low-WMC (Lunner, 2003; Peelle, 2018; Pichora-Fuller et al, 2016). This relationship suggests that high-WMC informs better long-term memory connections, use of context clues, and more successful repair of degraded perceptual inputs. Additionally, fewer WMC limitations have been associated with lower subjective reports of effort when completing these auditory tasks.

Rudner, Lunner, Behrens, Thorén and Rönnberg (2012) collected subjective effort ratings following two speech recognition in noise tasks for aided hearing-impaired listeners. Along with their experiments, WMCs were measured via a RSPAN and letter-monitoring task. The first speech recognition experiment presented Dantale II sentences and the second presented Hagerman Swedish sentences; both experiments had steady-state and ICRA background noise conditions. Background noise was adjusted via an adaptive SNR paradigm for the first task and was set at a fixed level for the second task. Unsurprisingly, effort ratings from the first experiment were highest when SNRs were the poorest. Additionally, effort ratings were higher for both experiments in the presence of ICRA background noise indicating that the mere presence of this noise increased cognitive demands of the task. The most informative finding from both experiments was that listeners with higher WMCs had lower overall effort ratings across conditions. These conclusions, like others (Francis & Strauss, 2017; Pichora-Fuller et al., 2016; Peelle, 2018), corroborate Rönnberg et al.'s (2013) assumption that higher WMC results in less effort needed to complete complex auditory tasks; however, generalization of these results is cautioned as listening effort studies typically only employ speech recognition tasks in the presence of background noise.

While tasks targeting speech perception and/or processing are important as they mimic daily situations, a few conceptual problems accompany them as well as their interpretations. First, is an argument well beyond the scope of the current work but relevant enough to be worth mentioning: the relationship between language mechanisms and executive functions is unclear as certain language processes are often discussed as "modular" – arguing that they *may* operate independently under certain conditions (Bowers & Davis, 2004; Myachykov, Scheepers & Shtyrov, 2013; Newport, 2011). Second – and more relevant – these assessments evaluate performance under *cognitive* loads or demands. Even when influences from *perceptual* features such as accents, fundamental frequencies, and compression are explored, results may still be confounded by cognitive load as verbal inputs have been argued to gain automatic access to post-perceptual levels of processing while non-verbal stimuli must be recoded in a lower-level buffer (Baddeley, 2010; Francis, Bent, Schumaker, Love, & Silbert, 2021; Repovs & Baddeley, 2006; Rönnberg, Holmer & Rudner, 2019; Vachon, Marsh & Labonté, 2020). This seemingly consistent presence of *cognitive* load makes it difficult to dissociate how much of auditory selective attention is driven by available WMC or top-down resources versus *perceptual* bottom-up stimulus features.

The inability to holistically define characteristics of load during complex listening is problematic, as patterns of distractibility vary depending on the form of load imposed during a task (Lavie, 2010; Murphy, Spence, & Dalton, 2017). This even persists for cognitive loads as recent evidence has been provided arguing that increased demands of a central/focal task allow individuals to achieve higher task engagement and experience reduced distractibility which opposes longstanding impressions of cognitive load where distractibility is mostly increased (Halin, Marsh, Hellman, Hellström & Sörqvist, 2014; Sörqvist, Nöstl & Halin, 2012; Sörqvist & Rönnberg, 2014;). Additionally – outside of the auditory domain – increases in visual *perceptual* load has been shown *reduce* distractibility and steadfast accounts of this phenomenon have not been replicated in the auditory domain. In the next section we will briefly discuss evidence for both patterns of distraction observed under cognitive load, introduce accounts of visual perceptual load, and dive into controversies as well as remaining questions regarding their applications to the auditory domain.

Load and Distractibility

The purpose of the load theory of attention is to evaluate how task demands influence the efficiency and method (i.e., *early* versus *late*) of selective attention (Forster & Lavie, 2009; Lavie, 2010; Lavie, Hirst, de Focket & Viding, 2004; Lavie & Tsal, 1994; Spence & Santangelo, 2010). The load theory assumes two types of load: perceptual and/or cognitive. Perceptually loaded tasks require bottom-up judgements of stimuli, while cognitively loaded tasks require top-down processing. The previously discussed listening effort experiments have been centered around *cognitive* load as top-down repair of degraded speech was required (Rönnberg et al., 2013; Rönnberg, Holmer & Rudner, 2019) and complex background noise was included (DiGiovanni, Riffle, Weaver & Lynch, 2017; Meister, Rählmann & Walger, 2018; Rudner, Lunner, Behrens, Thorén &

Rönnberg, 2012; Sarampalis, Kalluri, Edwards & Hafter, 2009; Trimmel, Schätzer & Trimmel, 2014).

Per the cognitive load theory, high load results in increased distractibility and low load allows for mediation of distractors. Following conclusions from reports of listening effort (Lemke & Besser, 2016; Rönnberg et al., 2013; Peelle, 2018; Rudner, Lunner, Behrens, Thorén & Rönnberg, 2012), listeners with increased WMC had residual executive resources during speech understanding tasks. It can be inferred that cognitive load limits for listeners with higher WMC were not reached, while limits for low WMC listeners were. Operating at limits of cognitive load may have resulted in increased reports of effort; however, this cannot be confirmed in this specific context as load studies predominantly assess interference from singleton distractions rather than reports of effort.

Additionally, factors affecting the generalizability of this claim stem from arguments and results suggesting that *increased* working memory load instead *reduces* the ability to perceive distractors (Halin, Marsh, Hellman, Hellström & Sörqvist, 2014; Simon, Tusch, Holcomb & Daffner, 2016). While Sörqvist, Stenfelt and Rönnberg's (2012) discussion was presented earlier in this review, a major part of their findings was withheld for this portion of the discussion. If you recall, the objective of their task was to determine what happens to wave-V of the ABR in the presence of cognitive load and determine whether individual capacity limitations (i.e., WMC) differences contributed to the ability to suppress responses to distracting information. On one hand, the results supported the controlled attention view in that lower capacities were more likely to experience intrusions of distraction reflected by larger ABR wave-V amplitudes. The second portion introduced here indicated that poorer ABR wave-V amplitudes, although still larger for low-capacity listeners, were modulated by task load; greater working memory loads resulted in reduced distractor processing than conditions where working memory load was less (again – measured via changes in "n" in an n-back where increased "n" equals higher load). From these findings as well as similar studies, spawned the notion that increased memory loads induce *engagement,* and greater *engagement* reduces perception of irrelevant stimuli introduced to tasks (Halin, Marsh, Hellman, Hellström & Sörqvist, 2014; Sörqvist & Marsh, 2015; Sörqvist & Rönnberg, 2014).

Like the latter findings, and contrary to typical assumptions of cognitive load, the perceptual load theory argues for increased distractibility under low load and reduced distractibility under high load (Lavie, 2010; Lavie & Tsal, 1994). Appropriate perceptually loaded tasks should not target functions such as working memory and inhibition, instead selective attention should be the primary function engaged. Distractibility under high load is reduced because attentional resources are constrained by target perceptual features and irrelevant stimuli do not gain access to focal attention (Broadbent, 1958; Driver, 2001; Lavie & Tsal, 1994; Spence & Santangelo, 2010; Treisman, 1964). Increased distractibility is observed under low load because irrelevant stimuli can gain access to focal attention and must be acted on to reduce further interference with performance. The most common manipulation of perceptual load has been confined to visual variations of flanker paradigms (Erikson & Erikson, 1974; Lavie, 2010; Lavie, Hirst, de Fockert & Viding, 2004; Murphy, Spence & Dalton, 2017).

Two adjustments of perceptual load in these flanker tasks are: 1) increase (high load) or decrease (low load) the number of items in a central array surrounding a target, or 2) increase (high load) or decrease (low load) the level of similarity between the array and target items (Driver, 2001; Gomes, Barrett, Duff, Barnhardt & Ritter, 2008; Lavie, 2010; Lavie, Hirst, de Fockert & Viding, 2004; Murphy, Groeger & Greene, 2016; Murphy, Spence & Dalton, 2017). Both increasing the "set-size" as well as "similarity" between array and target stimuli, result in more complex visual search fields, ultimately reducing the ease of target-object identification which I have adopted as a definition of perceptual load. Under low load, simplicity of the visual search array does not confine resources and allows for focal attention to extend beyond the array and perceive response competing (i.e., incongruent/incompatible) distracting objects. Under high load, target-distractor relationships should matter less as the objects outside of the central array are not perceived at higher levels (Lavie, 2010; Lavie, Hirst, de Fockert & Viding, 2004; Murphy, Spence & Dalton, 2017).

Researchers have attempted to adjust auditory perceptual load; however, data have not resulted in clear patterns suggestive of successful manipulation (Gomes, Barrett, Duff, Barnhardt & Ritter, 2008; Fairnie, Moore & Remington, 2016; Francis, 2010; Murphy, Fraenkel & Dalton, 2013; Strauss & Francis, 2017). Murphy, Spence, and Dalton (2017) conducted a review discussing these attempts and divided them into three overarching categories: 1) *number of items in the display,* 2) *similarity between non-targets and targets*, and 3) *number of perceptual operations required by a task*. Altering timing between presentations of stimuli (i.e., inter-stimulus-intervals (ISI)) was a common load manipulation across experiments; lower load allowed more time between stimuli and load was increased by reducing timings. The reviewed studies measured distractor interference via objective electrophysiologic measures of N1 (Parasuraman, 1980; Murphy, Spence & Dalton, 2017) and MMN (Woldorff et al., 1991; Murphy, Spence & Dalton, 2017) response amplitudes. For both experiments, reduced ISI (High-Load) resulted in smaller neural response amplitudes compared to those collected under Low-Load. These results were not conclusive though as they did not provide measures of behavioral interference and were not replicated in a later study by Gomes et al. (2008) who did not find any effects of ISI manipulation (Murphy, Spence & Dalton, 2017). The biggest questions resulting from these reviewed studies were: 1) what stimulus features (i.e., tones, noise-bursts, environmental, speech, etc.) as well as 2) experimental auditory paradigms result in a true unclouded *perceptual* load.

To investigate these questions further and following the trend of ISI manipulations in suggested auditory perceptual load work, I recently conducted a study prior to the current work designed to assess behavioral effects of ISI changes during a dichotic listening task. For that particular study ISIs are better described as "inter-trial-intervals" (ITIs) as they were defined by timing differences allowed between judgements of target sounds. Participants were instructed to determine if target spoken digits (one through nine) presented in their right ear were "odd" or "even", while ignoring simultaneous distractor sounds in their left ear. Distractors were either a standard (80% of trials) 440 Hz tone or deviant (20% of trials) white-noise burst. Under Low-Load a 500ms silent interval was added after responses were given (i.e., between trials), while

High-Load was defined by the absence of any silent interval (i.e., immediately after a response, the next trial's sound was introduced). Overall, ITI load manipulations had no effect on distractor interference as patterns did not differ between High- and Low-Load conditions (Lynch & DiGiovanni, 2019). Initial interpretations of these findings suggested the inability to translate patterns of distraction observed during visual perceptual load to the auditory domain; however, *perceptual* load manipulation may have been confounded by the task of labeling spoken digits as "odd" or "even" eliciting higher-level *verbal* processing and/or identifying the right ear as the target ear for each block during the dichotic listening task. The consistent use of the right ear as the "target ear" may have allowed listeners to "build-up" strategies directing or *lateralizing* attention towards the target ear and against the perception of sounds in the non-target, which occurs independently of task loads or target stimulus types (Bookbinder & Osman, 1979; Pollmann, Maertens, von Cramon, Lepsien & Hugdahl, 2002; Dos Santos Sequeira, Specht, Hamalainen & Hugdahl, 2008).

Noted by Murphy, Spence and Dalton (2017), perceptual load theory has mostly been retrospectively applied to describe performance on listening tasks rather than manipulated directly with the exception of a few studies (Francis, 2010; Fairnie, Moore & Remington, 2016; Gomes, Barrett, Duff, Barnhardt & Ritter, 2008; Murphy, Fraenkel & Dalton, 2013). Reverting back to the discussions of the "cocktail party effect" and *early* versus *late* or *attenuation* models of selective attention - it was assumed that the *perceptual* demands of target-talker voice features resulted in *early* selection and inhibited the ability to perceive information in the non-target ear (Murphy, Spence &

Dalton, 2017). Findings from Lynch and DiGiovanni (2019), however, may indicate that it was higher level engagement or processing of target speech in these "cocktail party" designs that reduced interference which would instead be supported by views of increased cognitive load "shielding" against distraction (Halin, Marsh, Hellman, Hellström & Sörqvist, 2014; Sörqvist & Marsh, 2015; Sörqvist & Rönnberg, 2014).

This lack of explicit manipulation as well as high variability between forms of auditory perceptual load and distractor paradigms across the reviewed studies have inhibited the ability to arrive at definite conclusion of the applicability to the auditory domain, an issue that is commonly noted at the end of each similar auditory perceptual load study as the work presented here (Francis, 2010; Gomes, Barrett, Duff, Barnhardt & Ritter, 2008; Murphy, Fraenkel & Dalton, 2013). Additionally, perceptual load theories assume that individuals have limited perceptual capacities. The question still remains regarding whether or not the lack of auditory perceptual load replicability results from the seemingly limitless perceptual capacity of auditory senses compared to visual, which can be reached by bottom-up load manipulations and cannot be in the auditory domain (Gomes, Barrett, Duff, Barnhardt & Ritter, 2008; Murphy, Fraenkel & Dalton, 2013; Murphy, Spence & Dalton, 2017). Furthermore, original visual load designs argue for close relationships between perceptual capacities and focal attention, perhaps even attentional capacities in general; however, measures of WMCs are often left out of these discussions especially in the auditory domain. Influence from WMC is worth investigating considering their contributions and insights to the size as well as efficiency of focal attention (Cowan, 2010; McElree, 2001; Oberauer, 2019; Shipstead, Harrison & Engle, 2015).

Motivation for the Current Work

Despite the absence of numerous reports, there is still evidence that suggests predictions of perceptual load theories can be observed in the auditory domain (Fairnie, Moore & Remington, 2016; Murphy & Greene, 2017; Murphy, Spence & Dalton, 2017; Sörqvist, Nöstl & Halin, 2012). Regardless of whether exact perceptual load theory predictions transfer to audition, additional investigations are warranted as bottom-up auditory demands must influence selective attention and interact with daily central listening demands in some fashion (Francis, 2010; Gomes, Barrett, Duff, Barnhardt & Ritter, 2008). To address the current gaps in related work, the present study was comprised of two listening tasks each with direct High- and Low- perceptual load manipulations, High- and Low- WMC listener groups, and three background noise conditions (i.e., Quiet, Steady-state noise [SSN], and multi-talker babble [MTB]).

Experiment I was designed as a dichotic listing task as these paradigms are traditionally employed to investigate selective attention and direct manipulations of perceptual load are lacking (Bookbinder & Osman, 1979; Driver, 2001; Cherry, 1953; Murphy, Fraenkel & Dalton, 2013; Murphy, Spence & Dalton, 2017; Moray, 1959). To control for potential lateralization of attention towards the target ear and away from the non-target, the target ear was switched between each block of trials; thereby eliminating the time allowed to develop this "build-up". Perceptual load was modulated via changing the number of characteristics or features needed to be perceived to judge a particular sound as a target (Alain & Izenberg, 2003; Murphy, Fraenkel & Dalton, 2013; Murphy, Spence & Dalton, 2017). Under Low-Load, listeners provided responses determined by which sound they heard (i.e., tone or noise-burst), while High-Load required judgements of both which sound was heard and if it aligned with a designated duration. This load manipulation was adapted from Alain and Izenberg's (2003) study where under Low-Load stimulus duration deemed a sound as a target and under High-Load both tuning and duration became important (Alain & Izenberg, 2003; Murphy, Spence & Dalton, 2017). One issue with Alain and Izenberg's (2003) study with respect to perceptual load is that higher-order inhibitory mechanisms may have been recruited since responses were *only* given if a target was heard in the directed ear – meaning other sounds that were not targets were also heard in the directed ear. Withholding responses may have elicited a "Go-No Go" type inhibition task breaking the intended perceptual nature of the task (Tiego, Testa, Bellgrove, Pantellis & Whittle, 2018). To prevent "Go- No Go" inhibitory control from occurring in Experiment I, each sound heard in the designated target ear was paired with a response. Additionally, Alain and Izenberg (2003) presented simultaneous oddball distractors to the non-target ear. Staying true to the origins of the load theory, a response competition paradigm was implemented here within Experiment I. This potentially adds another load aspect noted by Murphy, Spence and Dalton (2017) and included within Murphy, Fraenkel and Dalton's (2013) study investigating degrees of similarity between targets and distracting sounds during perceptual load tasks.

A response competition distractor paradigm was also implemented for Experiment II; however, instead of dichotic listening Experiment II assessed diotic listening where sounds are available to both ears rather than to each ear separately. Not only does this type of task have more overlap with real-world listening scenarios, it was modified from Fairnie, Moore and Remington's (2016) design that provided strong support for the perceptual load theory in the auditory domain. Within their study, environmental sounds were presented from up to six different spatial locations (1- only the target, 2- target plus array item, 4, and 6 were potential "set sizes"). Their overarching goal was to determine whether the increase in array items impaired the ability to perceive (i.e., awareness of) a non-target car sound (e.g. *critical sound*) that was presented beyond the central array. Fairnie, Moore and Remington (2016) concluded that increases in load via set size adjustments reduced the ability to notice the critical sound. While results clearly support this claim, other researchers have suggested that requiring participants to respond to distracting stimuli inadvertently turns the distracting object into a secondary target. Including a secondary target is similar to adding a dual-task; therefore, reduced responses to the critical sound may instead reflect a dual-task trade-off (Murphy, Spence & Dalton, 2017; Sörqvist & Rönnberg, 2014). A benefit of the response competition paradigm employed for Experiment II compared to this awareness report is that it avoids this potential dual-task confound, and again assesses a the factor of perceptual similarity between targets and distractor sounds.

While perceptual loads of Experiments I and II resolve some consistency issues that have impaired cross-task comparisons resulting from the use of various distractor paradigms as well as incongruous load manipulations, replicability of perceptual load in the auditory domain was only one aim of the current work. The broader scope was to investigate the ecological validity of perceptual loads for listeners since we all encounter complex demands in environments with both perceptual and cognitive traits. The most common *cognitive* factor is background noise, and negative effects are not only observed when levels reach those of intended inputs (i.e., reduced signal-to-noise ratios (SNRs)). Meister, Rählmann and Walger (2018) presented older listeners with low level speech shaped noise and assessed sentence recognition abilities. Even at these "favorable" noise levels, sentence recognition was impaired. The addition of cognitive load from this noise was assumed, as higher WMC listeners performed better than those with low-WMC. DiGiovanni, Riffle, Lynch and Nagaraj (2017) found similar results within both younger and older populations where background noise impaired attention switching and working memory running span performance regardless of that noise type. There is ample evidence supporting the notion that competing background sounds contribute to increased processing demands, but – noted previously – these effects are constrained to higher-order tasks and have not been applied to perceptually loaded tasks (Herwig & Bunzeck, 2015); therefore, it seems reasonable to question whether background sounds simply *add* to cognitive loads or *induce* cognitive loads when they are more or less absent. In the pursuit of ecological validity, WMCs have been included as a listener trait that may provide information regarding their relationships to auditory perceptual capacities and assist with the ability to determine the nature of the load imposed across Experiments I and II.

Given the relevance and wide variability of auditory distractors that are encountered during daily listening, better understandings of auditory selective attention have become necessary (Shinn-Cunningham & Best, 2008). Currently, reduced interference from irrelevant information has been attributed to available executive function resources, larger WMCs, and increased engagement (Halin, Marsh, Hellman Hellström & Sörqvist, 2014; Sörqvist & Marsh, 2015; Sörqvist & Rönnberg, 2014). A commonality among these factors is their association to top-down processing. Top-down engagement is often effortful and suggests that *late* or *attenuation* methods of selective attention are employed in the presence of irrelevant auditory inputs. Should patterns from visual perceptual load be observed in the auditory domain, evidence for *early* and *effortless* mediation of distractors that is driven by lower level (i.e., bottom-up) demands would be provided. Following difficulties highlighted within listening effort literature, environmental factors that may reduce effort are of importance. This is not to say that results unsupportive of the transferability of perceptual load theory are insignificant; instead, this outcome would add an additional consideration when analyzing specific listener complaints, their environments, and paradigms developed by researchers. The two experiments of the current study have been designed to minimize contributions from additional executive functions to paint a clearer picture of selective attention in the face of auditory distractors.

Specific Aims and Hypotheses

The aims across both Experiments I and II were to: 1) investigate the applicability of perceptual load to the auditory domain, 2) determine whether the addition of background noise compared to quiet creates cognitive load in the absence of cognitive task demands, and 3) provide novel insights regarding the relationship between WMCs and auditory perceptual loads. The collective objective of these three aims was to provide novel data that can aid the ability to predict instances of listener distractibility and continue disentangling both external and intrinsic factors that add to overall listener complaints in complex scenarios. To obtain an unconfounded viewpoint of bottom-up perceptual effects, data was collected from young-normal hearing listeners to eliminate signal distortion and all target sounds were non-verbal. Hypotheses one and three (H1, H3) were derived from expected findings from visual and cross-modal perceptual load findings. Hypotheses two and four (H2, H4) were inferred from related cognitive hearing science and listening effort work.

H1: Increased auditory perceptual load for both experiments will result in increased RTs and higher error rates.

H2: The introduction of background noise will result in patterns reflective of cognitive load.

H3: Effects of incongruent or response competing distractors will be observed only under Low-Load conditions.

H4: WMC-groups will not be influenced differently by changes in load, since *perceptual* demands influence bottom-up processes and limit engagement of executive functions (i.e., top-down processes).

Chapter 3: Methods

General Methods

Two versions of the tasks for Experiments I and II were developed: 1) "In-Lab" and 2) "Remote". The "In-Lab" version was programmed with E-Prime 3.0 (Psychology Software Tools, Sharpsburg, PA) and then converted to the "Remote" format through E-Prime Go (Psychology Software Tools, Sharpsburg, PA) to allow for data collection following the onset of the COVID-19 pandemic and university campus closures.

Participants

Thirty participants (Males = 6, Females = 24, *M-age* = 23.76 years, *SD-age* = 2.63) performed Experiments I and II. This sample size was derived from both an a priori power analysis generated with mean reaction times (RTs) from previous work (Lynch & DiGiovanni, 2019) and comparisons to samples used in related literature (Dalton, Santangelo & Spence, 2009; Fairnie, Moore & Remington, 2016; Simon, Tusch, Holcomb & Daffner, 2016). Seven of the thirty (Males = 1, Females = 6, *M-age* = 22.03 years, *SD-age* = 2.58) completed the "In-Lab" version, while the remaining twenty-three (Males = 5, Females = 18, *M-age* = 25.52 years, *SD-age* = 2.711) completed the "Remote" version. All participants passed a hearing sensitivity screening (0.5-4kHz) which was programmed through PennController for IBEX (pcibex.net; Zehr & Schwarz, 2018) for "Remote" participants as well as a Six-Item Cognitive Impairment Test (Callahan, Unverzagat, Hui, Perkins & Hendrie, 2002). Informed consent and screening procedures were held via Zoom conferencing for "Remote" participants and during a scheduled lab visit for "In-Lab" participants.

WMC

The Woodcock-Johnson-III: Auditory Working Memory test (Woodcock, McGrew & Mather, 2001; Schrank, 2010) served as a measure of WMC for all participants. This complex listening span assessment presents listeners with lists of spoken objects and digits; the task is to reorder those lists at recall. Listeners were specifically asked to verbally recall all of the objects in the order they were heard first followed by the digits. An example of a given trial may be: *Listen: "seven, pear"* to which correct recall would be "*pear, seven*". List lengths began short with only two-items and progressed up to lengths of eight; there were three trials at each length. Scoring for listeners persisted until two errors at a given list length were made.

WMC values were then calculated from the total number of correct trials, leaving a potential range from 1 (the lowest) to 21 (the highest). The WMC range for the total thirty participants was 8 to 19 ($M\text{-}wmc$ = 13, SD = 2.805), 8 to 19 for the seven "In-Lab" participants ($M\text{-}wmc$ = 13.86, SD = 3.441), and 9 to 19 for "Remote" participants ($M\text{-}wmc$ = 12.78, SD = 2.519). Listeners were separated into High- and Low-WMC groups via a median split procedure (Garofalo, Battaglia & di Pellegrino, 2019; Sörqvist, Nöstl & Halin, 2012). Following this method, Low-WMC listeners were defined by scores less than or equal to 13 (Total: Low-WMC n = 17, "In-Lab": Low-WMC n = 3, "Remote": Low-WMC n = 13), while High-WMC listeners scored above 13 (Total: High-WMC n = 13, "In-Lab": High-WMC n = 4, "Remote": High-WMC n = 10) (See Figure 1).

Figure 1

Distribution of all thirty participants WMC scores. The vertical dotted line represents the median split point separating the Low- (13 and below) and High- (above 13) WMC groups.

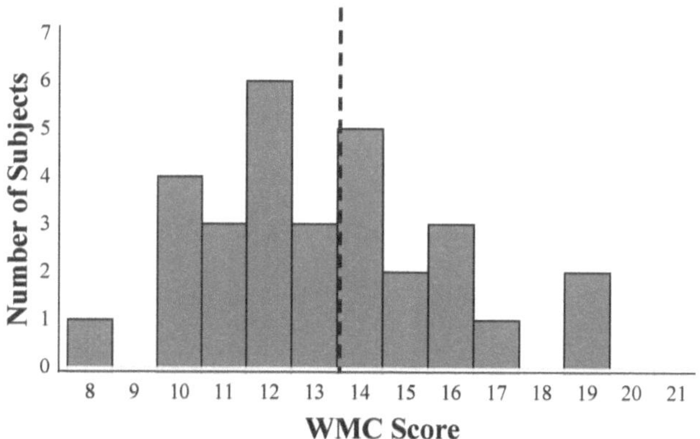

General Procedures

Both Experiments I and II consisted of two degrees of auditory perceptual load (High and Low) performed in three different background noise conditions (Quiet, SSN and MTB), for a total of six-conditions per experiment, or 12 total. Each condition was made up of four blocks; High-Load blocks of Experiment I as well as all blocks for Experiment II were comprised of 64-trials, while the Low-Load blocks of Experiment I had 60 trials. Background noise conditions within each experiment were counterbalanced to eliminate order effects. The total duration of participation had both Experiments I and II been performed successively, ranged from one hour and forty minutes to two and a half hours. Subjects who completed the tasks were compensated twenty dollars.

"In-Lab" participants completed all 12-conditions seated in a sound-treated booth in front of a computer monitor and keyboard, wearing Sennhieser HD 380 Pro headphones. "Remote" participants were instructed to find an *optimal* (i.e., quiet) listening setting to complete the tasks; this was surveyed over Zoom during the initial meeting. Given that "Remote" participant's environments were subject to change (i.e., a roommate coming home from work), they were encouraged to complete conditions separately rather than in one session should an unexpected intrusion occur. "Remote" participants were also instructed to use wired headphones to reduce potential delays and/or disconnections from Bluetooth headphones.

Additionally, a headphone orientation check was administered to "Remote" participants prior to data collection to ensure that: 1) listener headphones were correctly presenting separate sounds to each ear (i.e., Left and Right channels), and 2) the "target ear" was correctly identified – meaning subjects had the headphones placed on the correct ear in line with instructions for a given block. This headphone orientation check was again programmed through PennController (Zehr & Schwarz, 2018) and monitored during the initial Zoom conference.

Stimuli

To avoid potential effects from verbal processing of "target" sounds, stimuli used within Experiments I and II were either a 1kHz frequency modulated (FM) tone, white noise-burst, or environmental sound (i.e., dog's bark or duck's quack). The 1kHz FM tone and white noise-burst were generated through Adobe Audition (Adobe Systems Incorporated) and the environmental sounds were borrowed from the Soundsnap sound bank (Soundsnap, 2008-2021). Stimuli were then normalized to 65 dB SPL and set to durations of 200ms (B&K Type 2250 Sound Level Meter: Brüel & Kjær; Adobe Systems Incorporated); apart from the 1kHz FM tones and white noise-bursts used in High-Load conditions of Experiment I that had possible durations of 100 and 300ms which will be discussed later.

The goal of using 65 dB SPL was to achieve an *audible* and *comfortable* level for normal hearing listeners. While the exact 65 dB SPL level could not be guaranteed for "Remote" listeners, a loudness check programmed through PennController (Zehr & Schwarz, 2018) was administered and monitored via Zoom to achieve the same *audible* and *comfortable* criterion. During the loudness check, participants listened to four-second clips of SSN and MTB and were instructed to adjust their device volume settings to a comfortable level; participants could listen to each of these clips twice.

Additionally, SSN was generated through Adobe Audition (Adobe Systems Incorporated) and MTB was borrowed from the Words-in-Noise (WIN) test (Wilson, 2003). Background noise was set to a signal-to-noise ratio (SNR) of +10 dB, meaning that target and distractor stimuli were 10 dB louder than the background noise. Presenting background noise at 55 dB SPL should not induce any SNR loss for normal hearing listeners (Shadle, 2007; Le Prell & Clavier, 2017).

Distractor Paradigm

For all conditions of Experiments I and II, distractibility was induced using a congruency/response competition paradigm which is common for perceptual load designs (Lavie, 2010; Lavie, Hirst, de Fockert & Viding, 2004; Lavie & Tsal, 1994), where

distractor sounds share one of three relationships to a target on a given trial, congruent –
same as, incongruent – response competing, or neutral – no relationship; it is important to
note here that every trial had at least two potential target sounds. An example of this
paradigm under the following task instructions"…*decide whether you heard a TONE or
NOISE-BURST…*" with a *1kHz FM Tone* presented as a trial's target would be:
congruent = 1kHz FM tone, incongruent = white noise-burst, or neutral = dog's bark. All
three distractor types were equally likely to be encountered. Measures of distraction
(congruency effects) were calculated by the differences in RT performance between
congruent-neutral trials and incongruent trials.

Experiment I – Procedure

Experiment I was designed as a dichotic listening task. On each trial, participants
heard simultaneous target and distractor sounds presented to the right and left ears
separately. Instructions prior to experimental blocks informed the listener of the direction
of the "to-be-attended" target ear, which was alternated between blocks. Two out of the
four blocks presented targets to the right ear, while the other two presented targets to the
left ear (i.e., Block 1 = right ear, Block 2 = left ear, Block 3 = right ear, Block 4 = left
ear). Targets were 1kHz FM tones or white noise-bursts and participants provided
responses via keystroke on a keyboard. A tone corresponded to a response of "1" and a
noise-burst was paired with the "3" key. Participants had a maximum time of four-
seconds to provide a response prior to the termination of a trial; should participants
provide a response within four-seconds, the following trial was immediately presented.
Congruent distractors matched the 1kHz tone or white noise-burst target, incongruent

distractors were the other potential target sound, and neutral distractors were a duck's quack.

During the Low-Load condition, participants were only instructed to listen for which sound was heard (i.e., 1 kHz FM tone or white noise-burst set at fixed durations of 200ms). Low perceptual load was assumed as the features of the two target sounds are highly discernable and the identification of the target sound should only require bottom-up surveying of auditory stimuli.

To induce High-Load, durations of the sounds became important as well as which target was heard. Instead of both targets being set to 200ms, participants were instructed to respond "1" when there was a "short-100ms" tone, "3" for a "long-300ms" noise-burst, and "0" should the duration not match the block's instructions. Half of the blocks followed this "short-tone"/"long-noise" pairing, while the other half presented "long-tone"/"short-noise" target pairs. This paradigm assumes High-Load as an additional characteristic of each sound must be perceived before providing a response. Additionally, changing the duration of the target stimuli across blocks requires more complex perceptual operations.

Experiment II – Procedure

Instead of investigating selective attention through dichotic listening, Experiment II was designed to evaluate how auditory perceptual load influences spatial selective attention where listeners can use auditory cues from both ears to identify a target sound. To achieve spatial arrays in the absence of separate places in sound-field – or in other words, preserve head-related transfer functions (HRTFs) over headphones – auditory stimuli were recorded from different spatial locations in the following manner. Targets were always presented "in-front" of listeners, and distractors were presented to the "sides".

First, the "frontal" search array was centered around 0˚ azimuth or the space that aligns with a listener's nose. Four potential locations of horizontal "frontal" auditory stimuli were then separated to -18, -9, +9 and +18 relative to the 0˚ point in a straight line. These degrees of separation were selected to ensure each array item was perceived as a separate object. For normal hearing listeners, separation of 6˚ along a "frontal" horizontal plane is sufficient to achieve this (Carlile, Fox, Orchard-Mills, Leung & Alais, 2016). Refined localization abilities mattered less for Experiment II as participants were only reporting if a target was heard, they did not have to determine which location within the array it was presented from which also allowed for smaller degrees of separation between the array items. Two additional locations at +/- 90˚ were also set-up to allow for more extreme separation between distractor "flanker" sounds from the "frontal" search.

All sounds were recorded through Knowles' Electronic Manikin for Acoustic Research (KEMAR) (G.R.A.S. Sound and Vibration, Beaverton, OR) and angles of incidence were determined via arctan calculations; where the distance from the mid-point of the array was approximately 4.5 feet away from the center of KEMAR's artificial head and speaker cones were separated from each other by 8.5 inches. Environmental animal sounds (e.g. dog's bark and duck's quack), 1kHz FM tones, and white noise-bursts were recorded through KEMAR and a Roland R-26 (Roland Corporation, U.S., Los Angeles, CA) recorder from each potential "front" and "side" location independently. Individual sound (.wav) files were then combined in Adobe Audition (Adobe Systems Incorporated) according to specific task loads as well as distractor conditions.

Target sounds were always a duck's quack or a dog's bark similar to Fairnie, Moore and Remington's (2016) task that the present design was modeled after. Participants responded with a keystroke of "1" when they heard a *quack*, and "3" when they heard a *bark*. Perceptual load was determined by the number of other sounds included within the "frontal" array. Under Low-Load, a target was presented with one other sound making the array set-size a total of two items. High-Load was increased by adding three other sounds surrounding the target, resulting in an array of four items. Filler sounds included in both Low- and High- Load conditions were 1kHz FM tones or white noise-bursts to increase perceptual separation between the target and array items; thereby clarifying that the perceptual load manipulation was imposed solely by the size of the search rather than similarity factors. Like Experiment I, participants were allotted a maximum of four-seconds to provide a response before the next trial was automatically presented.

Chapter 4: Results

Results for Experiments I and II will be discussed in the order that the four hypotheses were presented in: 1) overall success of load manipulation, 2) effects of the addition of background noise, 3) measures of distraction, and 4) influence of WMC-group. Both accuracy and RTs served as outcome measures for analyses investigating success of load manipulation and overall task performance between the WMC-groups; however, only RTs were included within analyses assessing background noise and magnitudes of distractibility. Accuracy was excluded from the latter analyses as the nature of these perceptually loaded tasks were not exceptionally difficult, resulting in higher accuracy percentages near ceiling for all conditions. The use of RTs as the primary outcome measure provided insights to the timing of listener's abilities to develop perceptual judgements necessary to arrive at correct responses. Since Experiments I and II targeted selective attention, RTs also reflected direct influence from response competing distractors as well as increases in load on the efficiency of this mechanism; faster RTs and reduced RT differences between conditions are indicative of greater efficiency.

Manipulation of Load

A repeated-measures ANOVA was run to determine whether the perceptual load manipulations imposed by Experiment I (High-Load: comparison of four-perceptual features vs. Low-Load: 2-perceptual features) and Experiment II (High-Load: "frontal"-array with four potential locations vs. Low-Load: "frontal" array with two potential locations) were successful across both accuracy (proportion correct) and RT measures.

Paired-samples t-tests were then conducted to determine whether effects of load were consistent between "In-Lab" and "Remote" sub-groups to allow for the combination of these groups for subsequent analyses investigating effects of background noise conditions (Quiet, SSN, and MTB), measured distractibility, as well as WMC-group (High vs. Low). Results from accuracy analyses for Experiments I and II will be discussed first.

The omnibus test analyzing accuracy for Experiment I showed significant effects of perceptual load $F(1, 62) = 3329.092$, $p < 0.001$ and $F(1, 206) = 6331.358$, $p < 0.001$, for the "In-Lab" and "Remote" sub-groups, respectively. Since higher perceptual load resulted in reduced distractibility for both sub-groups, change in accuracy (Δ-Accuracy) was calculated to compare effects of load across these groups with the following equation: [Δ-Accuracy = Low-Load-MAccuracy – High-Load-MAccuracy]. A paired-samples t-test was then conducted to compare effects of load on accuracy between "In-Lab" "In-Lab" (M Δ-Accuracy = 0.1576, SD=0.181) and "Remote" (M Δ-Accuracy = 0.095, SD=0.1886) collection; results showed no differences in effects of load across sub-groups, $t(8) = -2.017$, $p = 0.078$. When the sub-groups were collapsed, accuracy remained poorer under High-Load (M= 0.78, SD=0.187) than Low-Load (M=0.91, SD=0.134) for Experiment I, $F(1, 268) = 92.861$, $p < 0.001$.

Similar results were obtained from the omnibus tests analyzing accuracy for Experiment II; $F(1, 62) = 15.141$, $p < 0.001$ and $F(1, 206) = 32.468$, $p < 0.001$, for the "In-Lab" and "Remote" subgroups, respectively. The paired-samples t-test for Δ-Accuracy between sub-groups ("In-Lab": M Δ-Accuracy = 0.0667, SD=0.068; "Remote": M Δ-Accuracy = 0.0317, SD=0.01472) was not significant, $t(6)=1.490$, $p = .196$. Accuracy for

Experiment II was again remained poorer under High-Load (M=0.8849, SD=0.134) than Low-Load (M=0.9249, SD=0.102) when the sub-groups were collapsed, $F(1, 269)$ = 46.887, p <0.001.

While High-Load conditions of Experiments I and II impaired overall accuracy, RT analyses were necessary to determine whether load manipulations were overall successful in line with visual perceptual predictions. Only RTs from correct responses that were greater than 200ms or less than 3 standard deviations (SD) from the mean were included within final analyses. Correct responses with these parameters were of primary interest as they suggest that participants were successful in judging perceptual features of targets and did not respond reflexively arriving at a correct response by chance, nor perform too slow on a rare outlying occasion. Percentages of removed data were minimal across Low- and High- conditions of Experiments I and II (4%, 3%, 3%, and 3%, respectively).

The omnibus test for Experiment I showed significant effects of load manipulation, $F(1, 61)$ = 39.634, $p < 0.001$ and $F(1, 206)$ = 240.140, $p < 0.001$, on RTs for the "In-Lab" and "Remote" sub-groups, respectively. Following the rationale and procedure used to analyze accuracy between sub-groups, ΔRT-Load was calculated with the following equation: [(High-Load mean RT) – (Low-Load mean RT)]. This ΔRT-Load served a secondary purpose as it controlled for unknown baseline timing differences between the E-Prime 3.0 and E-Prime Go platforms (Psychology Software Tools, Sharpsburg, PA) as well as potential differences of computer processing between "Remote" participants. The paired-samples t-test comparing ΔRT-Load (M ΔRT-"In-

Lab" = 187.88, *SD*=234.79; *M* ΔRT-"Remote" = 296.33, *SD*=275.05) between sub-groups was not significant, $t(8) = 1.578$, $p = 0.153$. With the two groups collapsed, RTs remained slower under Experiment I's High-Load ($M = 1002.12$, $SD = 210.596$) conditions compared to Low-Load ($M = 730.79$, $SD = 192.48$), $F(1, 268) = 272.058$, $p < 0.001$.

For Experiment II, the omnibus test again showed significant effects of load on RT for the "Remote" sub-group, $F(1, 206) = 11.358$, $p < 0.001$, but not the "In-Lab" sub-group, $F(1, 62) = 1.229$, $p = 0.272$; however, the paired-samples t-test comparing ΔRT-Load between High and Low-Load conditions for the groups (*M* ΔRT-"In-Lab" = 19.283, *SD*=47.62; *M* ΔRT-"Remote" = 17.167, *SD*=17.031) was not significant, $t(6) = 0.097$, $p = 0.926$. When both groups were collapsed, RT performance under High-Load ($M = 684.315$, $SD = 125.065$) was slower than Low-Load ($M = 665.97$, $SD = 127.075$) for Experiment II, $F(1, 269) = 8879.816$, $p < 0.001$ (See Table 1).

Table 1

Summary of paired-samples t-tests: Effects of perceptual load between "In-Lab" and "Remote" sub-groups.

Experiment	Sub-Group	Measure	Δ -Load (*M, SD*)	*t*	*p*
I	"Remote"	Accuracy	0.095, 0.189	-2.02	0.078
I	"In-Lab"	Accuracy	0.158, 0.181		
II	"Remote"	Accuracy	0.032, 0.014	1.49	0.196
II	"In-Lab"	Accuracy	0.066, 0.068		
I	"Remote"	RT	296.33, 275.05	1.578	0.153
I	"In-Lab"	RT	187.88, 234.79		
II	"Remote"	RT	17.167, 17.031	0.097	0.926
II	"In-Lab"	RT	19.283, 47.62		

While high perceptual load appeared to impose a significant effect on RTs across Experiments I and II, Δ-RT's were starkly different (Experiment I: Δ-RT = 271.33ms; Experiment II: Δ-RT = 18.345ms). To further investigate the success of load manipulations, Cohen's d's were calculated to compare the effect sizes of load between experiments. The effect size (Cohen's d) of load on mean RT was large for Experiment I (d = 1.125), yet minimal (d = 0.145) for Experiment II. Cohen's d for accuracy of Experiment I was also large (d = 0.765), while Experiment II was small (d = .347) (See Table 2).

Table 2

Effect size of perceptual load manipulation on Δ-RT and Δ-Accuracy for Experiments I and II. (Cohen's d).

Experiment	Measure	Δ -Load	d
I	Accuracy	0.13	0.765
II	Accuracy	0.045	0.345
I	RT	271.33	1.125
II	RT	18.345	0.145

Background Noise Conditions

High- and Low- perceptual load tasks were performed in three background noise conditions for Experiments I and II: Quiet, SSN and MTB. A univariate ANOVA was run for each experiment to determine whether the type of background condition influenced the amount of perceptual load imposed across tasks. ΔRT-Load's from the previous section were used as a measure of load.

Results from the omnibus test for Experiment I showed significant effects of type of background noise on the amount of load imposed, $F(2, 268) = 30.461$, $p < 0.001$. Post-hoc analyses (Bonferroni corrected) revealed significant differences in load during the Quiet conditions compared to both MTB ($p < 0.001$) and SSN ($p < 0.001$), but no differences between MTB and SSN ($p = 0.361$). These results indicate that effects of perceptual load were greatest in the absence of background noise ($M = 431.986$, $SD = 243.258$) and were reduced in the presence of background noise. The type of background noise introduced was irrelevant for changes in load (MTB: $M = 219.138$, $SD = 256.956$; SSN: $M = 162.291$, $SD = 232.145$) (See Figure 2).

Figure 2

*Influence of background noise condition on effects of perceptual load for Experiment I. Significant differences (p<0.05) are denoted with *.*

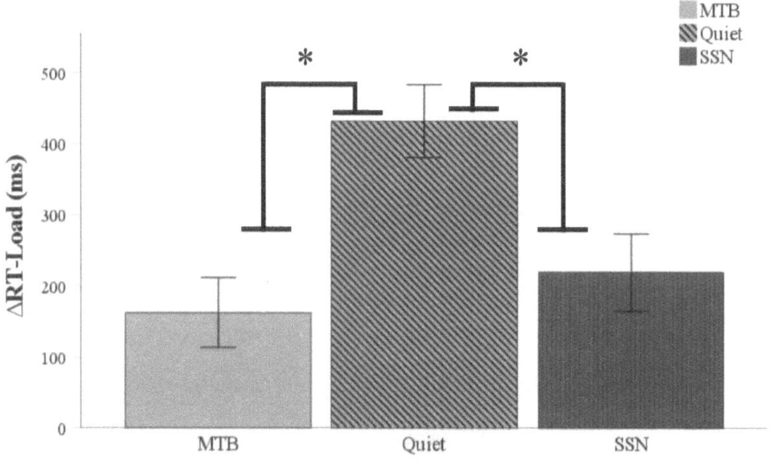

Results from Experiment II did not reveal any differences in effect of load across Quiet (M = 31.019, SD = 86.393), MTB (M = 7.181, SD = 86.393), or SSN (M = 20.305, SD = 106.644) background conditions, $F(2, 239)$ = 1.592, p = 0.206 (See Figure 3).

Figure 3

Influence of background noise condition on effects of perceptual load for Experiment II.

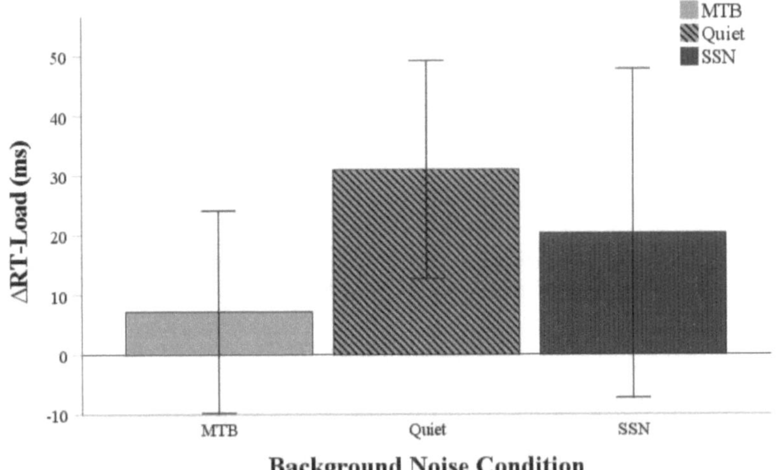

Distractibility

For both Experiment I and II, ΔRTs were calculated with the following equation to serve as a measure of distractibility: [ΔRT-Distraction = (IncongruentRT - (CongruentRT + NeutralRT)/2))]. Congruent and neutral distractor types were averaged into a composite value following the lack of significant differences between the two types (Experiment 1: Congruent-Neutral, p= 1.0; Experiment 2: Congruent-Neutral, p =0.188) and reports from related literature (Konstantinou, Beal, King & Lavie, 2014). A 2x3

ANOVA was then conducted to evaluate differences in ΔRT across load (High vs. Low)

and background noise condition (Quiet, SSN, MTB) for Experiments I and II.

The omnibus test for Experiment I showed a significant effect of load on ΔRT,

$F(1, 179) = 39.101$, $p < 0.001$; however effects of background noise conditions, $F(2, 179)$

$= 2.331$, $p = 0.100$, and the interaction between load and background noise, $F(2, 179) =$

0.506, $p = 0.604$, were not significant. Measured distractibility (i.e., ΔRT) was greater

under all High-Load conditions ($M = 144.439$, $SD = 123.88$) than Low-Load ($M =$

20.282, $SD = 143.066$) for Experiment I (See Figure 4).

Figure 4

*Experiment I ΔRT's by Load (High vs. Low) and background noise condition (MTB,
Quiet, SSN). Significant differences (p <0.05) are denoted with *.*

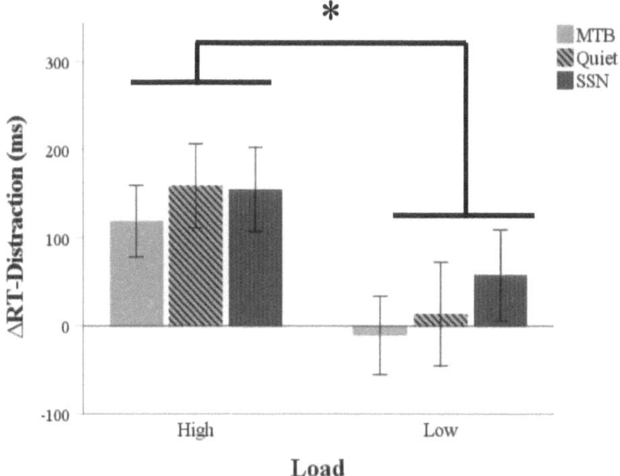

The omnibus test for Experiment II did not show significant effects of measured distractibility (i.e., ΔRT) across load [$F(1, 179) = 0.008$, $p = 0.927$], background noise condition [$F(2, 179) = 0.529$, $p = 0.590$], or their interaction [$F(2, 179) = 0.681$, $p = 0.507$] (See Figure 5).

Figure 5

Experiment II ΔRT's by perceptual load (High vs. Low) and background noise condition (MTB, Quiet, SSN).

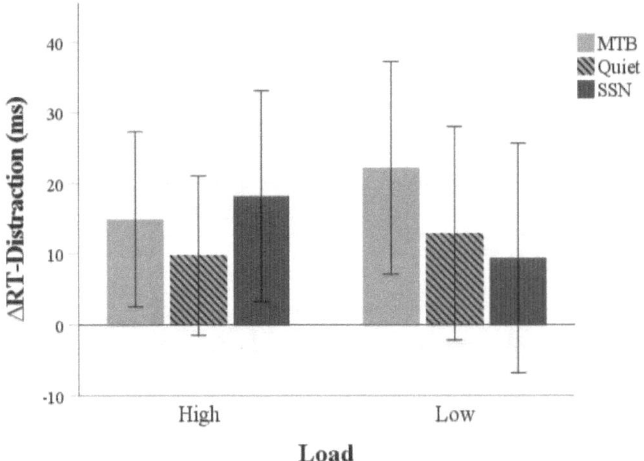

WMC Groups

To reiterate, WMCs were measured via the Woodcock Johnson III-Auditory Working Memory Test and participants were divided into High- and Low-WMC groups based on their total number of trials correct: High-WMC = greater than 13 trials, Low-

WMC = 13 trials or fewer. Analyses were run for Experiments I and II investigating whether WMC groups influenced overall effects of load (i.e., ΔRT-Load), measured distraction (i.e., ΔRT-Distraction), and overall task performance (i.e., mean RT and accuracy).

First, a 2x3 ANOVA was run for Experiment I to compare effects of load across WMC groups and background noise conditions. Results from the omnibus test showed that ΔRT-Load did not differ between WMC groups [$F(1, 268) = 0.032, p = 0.859$]. Consistent with results from Experiment I's analysis of background noise conditions, ΔRT-Load was influenced by noise-types, $F(2, 268) = 30.430, p < 0.001$. The interaction between WMC group and background noise condition was not significant [$F(2, 268) = 0.517, p = 0.597$], indicating that effects of background noise and load were consistent for High- and Low-WMC participants.

Again, post-hoc analyses showed increased effect of load in Quiet (High-WMC: $M = 447.551, SD = 179.054$; Low-WMC: $M = 420.083, SD = 283.957$) compared to MTB (High-WMC: $M = 140.761, SD = 172.068$; Low-WMC: $M = 178.754, SD = 296.790$) ($p < 0.001$) and SSN (High-WMC: $M = 234.075, SD = 239.959$; Low-WMC: $M = 207.488, SD = 271.307$) ($p < 0.001$), and no differences between MTB and SSN ($p = 0.366$) (See Figure 6).

Figure 6

Experiment I effects of load across WMC groups and background noise conditions.
*Significant differences (p <0.05) are denoted with *.*

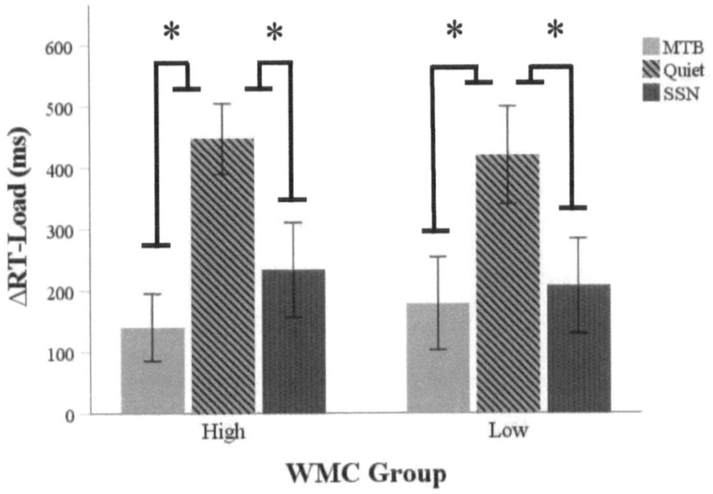

This was repeated for Experiment II. The omnibus test yielded significant effects of WMC group [$F(1, 239) = 4.498$, $p = 0.035$]; however, the test of background noise condition [$F(2, 239) = 1.541$, $p = 0.216$] and their interaction were both not significant [$F(2, 239) = 0.010$, $p = 0.990$]. Ultimately, High-WMC listeners ($M= 5.174$, $SD = 86.651$) were less affected by load manipulations in Experiment II than Low-WMC listeners ($M = 30.281$, $SD = 91.249$) (See Figure 7).

Figure 7

Experiment II effects of load across WMC groups and background noise conditions.
*Significant differences (p <0.05) are denoted with *.*

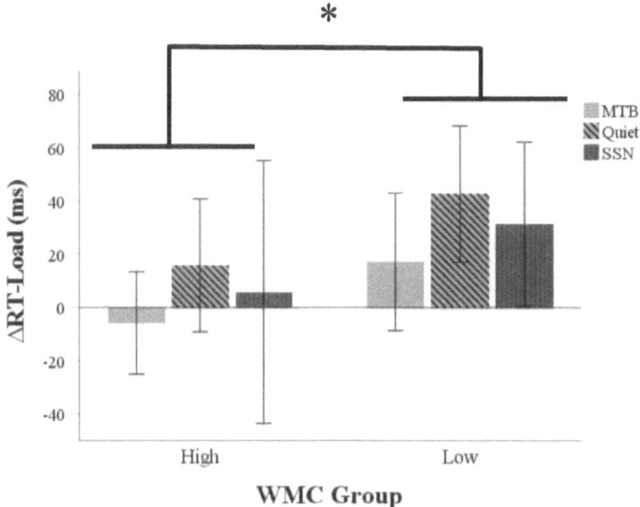

Following the analysis of load effects, a 2x2x3 ANOVA was run to determine

whether WMC Groups informed measured distractibility (ΔRT-Distraction) across

background noise conditions and load for Experiments I and II.

The omnibus test for Experiment I showed effects of load on measured

distractibility [$F(1, 179) = 35.608, p < 0.001$], but did not yield effects from WMC group

[$F(1, 179) = 0.014, p = 0.905$] nor background noise condition [$F(2, 179) = 1.832, p =$

0.163]. Two-way interactions between WMC group and load [$F(1, 179) = 1.512, p =$

0.221], WMC and background noise condition [$F(2, 179) = 0.168, p = 0.846$], and background noise condition and load [$F(2, 179) = 0.423, p = 0.650$] were all not significant. Additionally, the three-way interaction between WMC group, background noise condition, and load was also not significant, $F(2, 179) = 0.223, p = 0.800$ (See Figure 7). While distractibility was increased under High-Load, WMC group did not inform differences in distraction for Experiment I (See Table 3 for a summary and Figure 8).

Table 3

Experiment I summary of ΔRT-Distractibility by WMC-Group, Load, and background noise.

WMC Group	Load	Noise Condition	ΔRT-Distractibility (*M, SD)*
High	High	Quiet	134.12, 107.25
High	High	SSN	152.48, 103.29
High	High	MTB	99.87, 84.27
Low	High	Quiet	177.914, 146.06
Low	High	SSN	157.09, 150.32
Low	High	MTB	134.13, 140.26
High	Low	Quiet	16.83, 101.36
High	Low	SSN	64.69, 102.20
High	Low	MTB	17.15, 44.76
Low	Low	Quiet	11.28, 196.69
Low	Low	SSN	52.417, 168.47
Low	Low	MTB	31.76, 155.12

Figure 8

*Experiment I effects of distraction across WMC groups and background noise conditions. Significant differences (p<0.05) are denoted with *.*

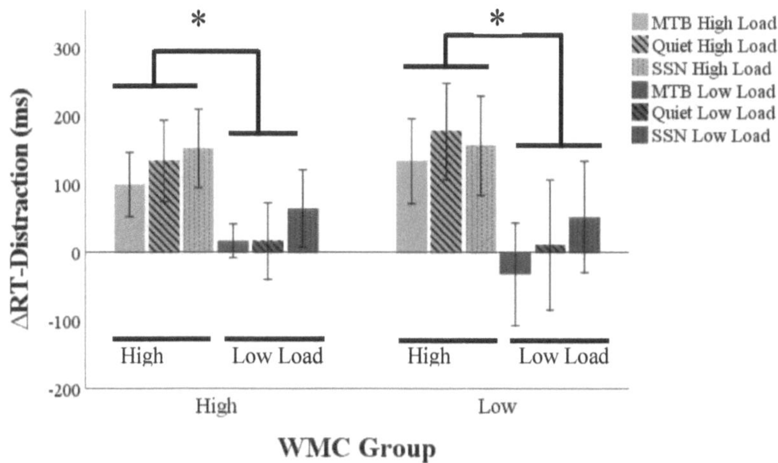

The omnibus test for Experiment II showed significant differences in measured distraction for WMC groups [$F(1, 179) = 19.244$, $p < 0.001$], but not across load manipulation [$F(1, 179) = 0.025$, $p = 0.874$] or background noise condition [$F(2, 179) = 0.480$, $p = 0.619$]. Two-way interactions between WMC group and load [$F(1, 179) = 0.227$, $p = 0.635$], WMC group and background noise condition [$F(2, 179) = 0.248$, $p = 0.781$], and background noise condition and load [$F(2, 179) = 0.734$, $p = 0.482$] were all insignificant. Additionally, the three-way interaction between WMC group, background noise condition and load was also insignificant, $F(2, 179) = 0.540$, $p = 0.584$. Ultimately, High-WMC listeners experienced greater distraction than Low-WMC listeners and effects of distraction were consistent across both High- and Low-Load conditions of

Experiment II (See Table 4 for a summary and Figure 9).

Table 4

Experiment II summary of ΔRT-Distractibility by WMC-Group, Load, and background noise.

WMC Group	Load	Noise Condition	ΔRT-Distractibility (*M, SD*)
High	High	Quiet	19.45, 36.88
High	High	SSN	33.07, 25.10
High	High	MTB	27.53, 35.08
Low	High	Quiet	2.45, 23.86
Low	High	SSN	6.83, 41.97
Low	High	MTB	5.34, 30.43
High	Low	Quiet	34.63, 40.96
High	Low	SSN	25.10, 43.26
High	Low	MTB	31.04, 42.77
Low	Low	Quiet	3.65, 34.01
Low	Low	SSN	2.51, 42.66
Low	Low	MTB	15.44, 39.43

Figure 9

*Experiment II effects of distraction across WMC groups and background noise conditions. Significant differences (p <0.05) are denoted with *.*

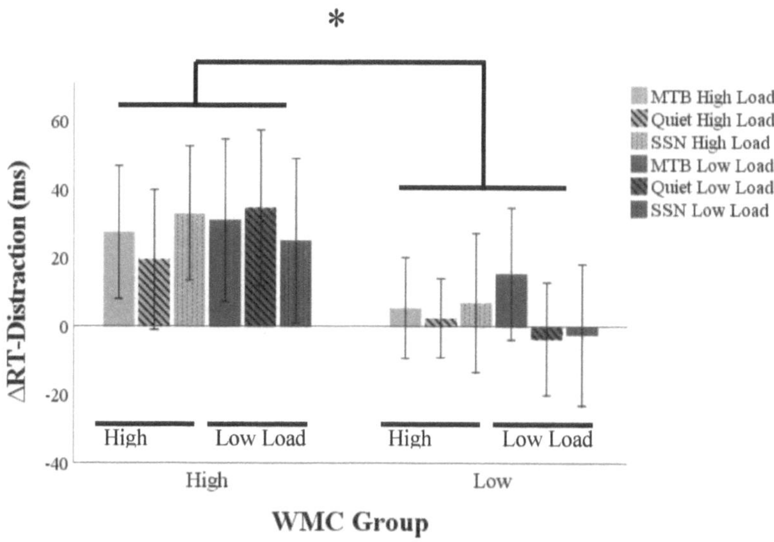

Final exploratory analyses were conducted to determine whether WMC group affected overall performance (i.e., mean RTs and accuracy) for Experiments I and II.

For Experiment I, Low-WMC listeners performed with poorer accuracy (M= 0.835, SD = 0.154) than those grouped with High-WMC (M = .880, SD = 0.124), $F(1, 179)$ = 4.531, p = 0.35, and also had slower mean RTs (Low-WMC: M = 905.268, SD = 219.253; High-WMC: M = 811.083, SD = 207.769), $F(1, 181)$ = 8.632, p = 0.004 (See Figure 10).

Figure 10

*Difference in mean RT (panel A) and accuracy (panel B) by WMC group for Experiment I. Significant differences (p <0.05) are denoted with *.*

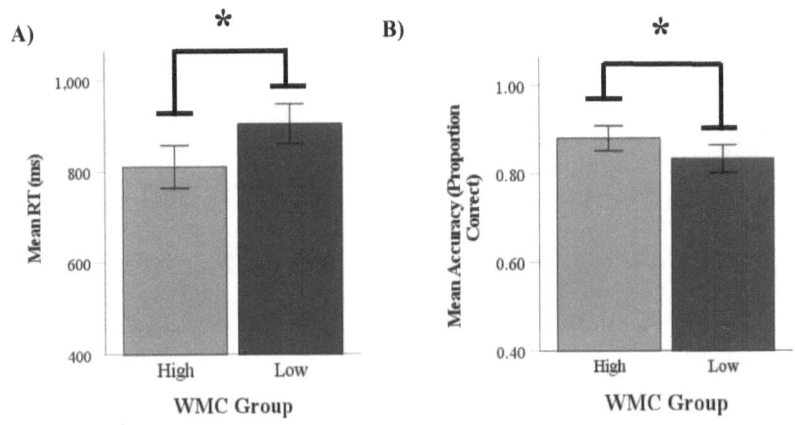

For Experiment II, the Low-WMC group again performed with poorer accuracy ($M = 0.886$, $SD = 0.033$) than the High-WMC group ($M = 0.929$, $SD = 0.078$), $F(1, 179) = 6.575$, $p = 0.011$, and slower mean RTs (Low-WMC: $M = 707.339$, $SD = 121.646$; High-WMC: $M = 633.038$, $SD = 110.650$), $F(1, 179) = 17.820$, $p < 0.001$ (See figure 11).

Figure 11

Difference in mean RT (panel A) and accuracy (panel B) by WMC group for Experiment II.

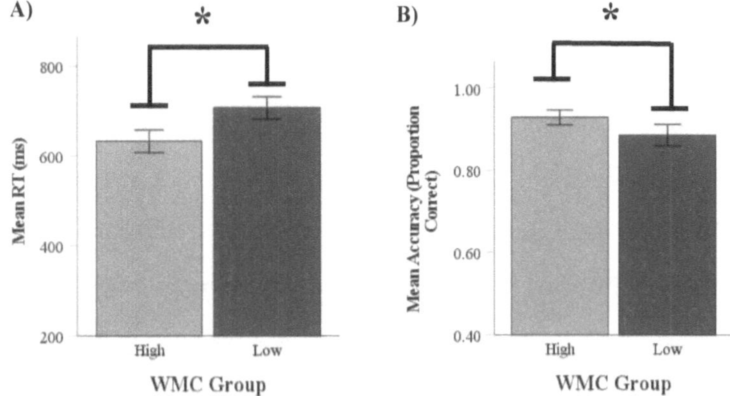

Chapter 5: Discussion

The overarching purpose of the current work was to provide a closer look at implications of perceptual load in the auditory domain, as perceptual load theory predictions predominantly stem from visual and/or cross-modal designs. The application of perceptual load theory to audition has perplexed researchers across multiple disciplines and created a divide among them as results have been very inconsistent. Part of this inconsistency stems from the lack of available studies investigating direct manipulations of auditory perceptual load. Another contributing factor is the disagreement over which manipulations are actually *perceptual* with respect to changes in complexities of sounds or auditory environments. Regardless, auditory perceptual load is becoming a "hot topic" within the realm of cognitive hearing science as the perceptual load theory is useful for describing patterns of selective attention (Francis, 2010; Murphy, Spence & Dalton, 2017). Having a greater understanding of auditory selective attention and how bottom-up perceptual demands influence its efficiency is necessary since this mechanism contributes to listeners' abilities to separate relevant from irrelevant or competing sounds (Shinn-Cunningham & Best, 2008). The perceptual load theory posits reduced distractibility when bottom-up demands are high, and increased distractibility when demands are low (Lavie, 2010; Lavie, Hirst, de Fockert & Viding, 2004; Lavie & Tsal, 1994); however, noted earlier on within this narrative, highly demanding scenarios are where listeners express the most difficulty.

Specific goals of the current study were to: 1) investigate influences from bottom-up perceptual auditory demands in isolation, and 2) include components that have been identified as contributors to complex listening performance to determine how they collectively interact with perceptual demands. To do so, Low- and High- manipulations of perceptual load were imposed across a dichotic (Experiment I) and spatial (Experiment II) listening task following recommended and/or suggested load manipulations from similar work (Fairnie, Moore & Remington, 2016; Francis, 2010; Murphy, Fraenkel & Dalton, 2013; Murphy, Spence & Dalton, 2017). To simulate more realistic listening scenarios beyond controlled laboratory environments and increase the ecological validity of the results, all conditions were performed in Quiet, SSN and MTB. Across both experiments, selective attention was gauged through measures of distraction to singleton sounds via a response competition paradigm. Participants were also separated into High- and Low-WMC groups to evaluate influences from listener characteristics, which has yet to be included within auditory perceptual load studies – at least at the time of the development of the current design.

This section is broken down into three parts where results from Experiments I and II will be discussed first followed by general conclusions. Primary analyses revolved around: 1) overall success of load manipulation, 2) influence of background noise condition (Quiet, SSN, and MTB), 3) changes in distractibility as a function of load, and finally 4) contributions from WMC-Groups.

Experiment I

The dichotic listening task designed for Experiment I altered perceptual load by changing the number of stimulus features to be compared across High- and Low-Load conditions, or *increasing the number of perceptual operations* as Murphy, Spence and

Dalton (2017) as well as Lavie (2010) would have categorized. During Low-Load trials, participants' objectives were to simply determine whether a 1kHz FM tone or white-noise burst was heard in a designated "to-be-attended" (i.e., target) ear. Under High-Load, participants had to judge both which sound was heard as well as if it matched the designated duration paired with that sound during the instruction set preceding an experimental block of trials (e.g.: *Target 1*= short 100ms – 1kHz FM tone, *Target 2* = long 300ms – white-noise burst). To prevent influence from potential strategies employed by listeners to effectively reduce perception of sounds presented to the non-target ear, or a "build-up" to ignoring sounds presented in the non-target ear (Bookbinder & Osman, 1979; Coleflesh & Conway, 2007), the to-be-attended ear switched between each block for both High- and Low-Load conditions of the task (e.g.: Block 1 = Right, Block 2 = Left, Block 3= Right, Block 4= Left). The first analysis was conducted to evaluate performance differences between High- and Low-Load conditions, ultimately assessing if manipulations were successful.

Per traditional perceptual load theory accounts, RTs should be longer and accuracy should be poorer under High- compared to Low-Load (Francis, 2010; Fairnie, Moore & Remington, 2016; Lavie, 1994, 2005, 2010; Lavie, Hirst, de Fockert & Viding, 2004). For Experiment I, these patterns replicated and supported the first hypothesis of the current work as mean RTs were longer and accuracy was poorer under the four-feature (two-sounds paired with two-durations) High-Load condition than the Low-Load two-feature (two-sounds) version of the same task. This increase in RT and reduction in accuracy indicates that the perceptual difficulty of the four-feature comparison task was greater than the two-feature identification task. While these findings align with visual perceptual load predictions, increased task difficulty regardless of type of demand or load (i.e., *cognitive* versus *perceptual*) result in similar patterns. In order to determine whether these effects truly aligned with perceptual load predictions, measures of distractibility were evaluated. Prior to diving into the discussion of measures of distractibility, it is important to review influence from the secondary load– the addition of background noise conditions.

Within most listening effort and cognitive hearing science studies, it has been argued that the addition of background noise induces some form of *cognitive* load. This is especially evident when the background noise is MTB or shares similar modulations to speech as the complexity results in greater attentional engagement to reduce interference from occurring (Pichora-Fuller et al, 2016; DiGiovanni, Riffle & Lynch, 2017; Meister, Rählmann & Walger, 2018; Tremblay, Brisson & Deschamps, 2021). One potential confound of this conclusion, is that effects of background noise have been primarily evaluated within tasks that already have a higher-level *cognitive* demand embedded (i.e., repair or recognition of speech, attention switching, etc.) and less is known about their impact on solely perceptually loaded tasks (Herweg & Bunzeck, 2015). The purpose of including background noise for the present study was to evaluate how the addition of MTB and SSN changed effects of load between High- and Low conditions or ΔRT-Load. It was expected that increased cognitive demands following the introduction of background noise would add to the demands of high-perceptual load resulting in larger ΔRT-Load values. Additionally, this increase in ΔRT-Load was anticipated to be greatest in the presence of MTB and would not differ between SSN and Quiet as stated by the second hypothesis.

Interestingly, ΔRT-Load's were actually *reduced* in the presence of both MTB and SSN compared to Quiet and did not differ between the noise types. This suggests that the presence of either type of background noise reduced the difference in RTs between High- and Low- conditions, or made RTs similar; in other words, effects of load were greatest in Quiet. That being said, ΔRT-Load values in MTB and SSN were still positive and significantly varied from zero, so it would be erroneous to conclude that the addition of background noise ameliorated effects of perceptual load all together; however, it is likely that low-level MTB and SSN increased the complexity of the low-load task which resulted in longer RTs closer to those measured under high-load. To get a better idea of whether the addition of background noise changed the nature of task load (i.e., *cognitive* versus *perceptual*) as well as establish which theory aligns with performance in Quiet, measures of distraction were evaluated next.

Changes in (Δ) RT-Distraction were calculated to serve as the measure of listener distractibility via subtracting the average composite RT of congruent-neutral trials from mean RTs during incongruent distractor trials. According to the perceptual load theory and hypothesis two of the current work, ΔRT-Distraction should be smaller under High- compared to Low-Load versions of the dichotic listening task. Contrary to expected patterns, ΔRT-Distraction was *greater* under High-Load and neared zero under Low-Load. Additionally, ΔRT-Distraction values did not differ under either High- or Low-Load between Quiet, SSN nor MTB. Given that measured distraction has been the primary determinant of load types, this lack of difference supports the notion that the addition of background noise did not change the nature of the load imposed by the task, or impose a significant additional *cognitive* demand.

The primary theoretical foundation driving the expected effects of reduced distractibility under high perceptual load is derived from the notion that individual's peripheral perceptual capacities are limited (Fairnie, Moore & Remington, 2016; Forster & Lavie, 2009; Francis, 2010; Murphy, Spence & Dalton, 2017). Reverting back to the discussion of accounts or methods of selective attention, *early* selection is assumed to accompany high perceptual demands. This assumption suggests that high demands absorb all peripheral capacity, thereby constraining focal attention to target-item identification and eliminating perception of non-target or distracting items (Lavie, 2010; Lavie, Hirst, de Fockert & Viding, 2004; Lavie & Tsal, 1994). The increase in distractibility observed here within Experiment I does not indicate that increased auditory perceptual demands promoted *early* methods of selection, and instead suggest that auditory selective attention operated *late* regardless of load. These patterns of distractibility may align with two alternative explanations.

First, the addition of features to be compared during the High-Load might have inadvertently induced a *cognitive* load. This interpretation is centered around a critique offered by Murphy, Spence and Dalton (2017) within their review discussing similar auditory perceptual load manipulations (Francis, 2010; Gomes, Barrett, Duff, Barnhardt & Ritter, 2008; Murphy, Fraenkel & Dalton, 2013). Studies discussed within their review, however, either resulted in no distractibility differences between High- and Low-load, were supportive of extensions of perceptual load theory to the auditory domain (i.e., reduced distractibility under high-load), or intentionally included a secondary dual-task; whereas the current study revealed patterns of distraction that are in direct opposition to perceptual theory accounts. Had a *cognitive* load been imposed within Experiment I, it would have resulted from the need to develop as well as retain "templates" in working memory that housed stimulus characteristics to listen for paired with correct button-responses (Bengson & Mangun, 2011; Coleflesh & Conway, 2007). This is not uncommon as most tasks require participants to form goal-directives based off instruction-sets, but may have become cognitively demanding here with the addition of finite stimulus features (i.e., short versus long and type of sound). The second potential explanation has been used in the past to explain why predictions of perceptual load theory are inconsistently observed within the auditory domain. Here, the argument is that limits of visual perception are much greater than those for audition, and therefore cannot be reached by increases in auditory perceptual load (Murphy, Fraenkel & Dalton, 2013; Murphy, Spence & Dalton, 2017). A benefit of the current work is that a measure of WMC was included which assists with the ability to interpret which type of load was imposed by the task, as well as whether auditory perceptual capacities can be reflected by this general measure.

Performance differences between WMC-groups are largely unknown with respect to isolated changes of perceptual loads or demands, especially within the auditory domain. That being said, expected WMC-group differences were derived from controlled attention views of WMC. If you recall from earlier sections, controlled attention views support the notion that WMCs reflect differences in flexible control over attentional resources as well as items in focal attention, where higher capacity individuals are more successful than those with lower capacities. This improved or "superior" task performance often exhibited by high-WMC individuals is predominantly observed when tasks require recruitment of top-down executive functions and become ameliorated when recruitment requirements are removed (Coleflesh & Conway, 2007; Engle, 2018; Heitz & Engle, 2007; Shipstead, Harrison & Engle, 2012). Given that the goal of *perceptually* loaded tasks are to limit contributions or engagement of higher order functions to determine how bottom-up features of objects influence selective attention - it was assumed that WMC-group differences would be absent. In other words, the fourth and final hypothesis states that high- and low-WMC listeners would be influenced by perceptual load manipulations of Experiment I similarly. Results align with these anticipated findings, as both ΔRT-Load and ΔRT-Distraction were not informed by WMC-group; however, when overall task performance was evaluated listeners categorized with high-WMC presented with faster mean RTs and better accuracy than those with low-WMC.

While the lack of WMC-group differences for ΔRT-Load and ΔRT-Distraction aligned with expected results, they add to the puzzling question of *why* distractibility was greater under *high-perceptual* load as well as what type of load was actually imposed by the increase in number of stimulus features to be compared (i.e., *number of perceptual operations*: Murphy, Spence & Dalton, 2017). The argument presented earlier on in this section suggested that the load manipulation of Experiment I induced a *cognitive* rather than *perceptual* demand. This *cognitive* load would have resulted from the need to develop and maintain complex "templates". The higher order demand of goal-directive retention would have led to increased distractibility as well as longer RTs and poorer accuracy according to typical cognitive load theory predictions; however, the lack of WMC-group differences challenges the plausibility of this interpretation.

Had this form of *cognitive* load been induced, high-WMC listeners would have been expected to present with reduced distractibility or smaller ΔRT-Distraction values that were closer to zero. This is a primary argument in support of controlled attention views of WMCs, as high-WMC listeners should be able to maintain goal directives in activated states longer and more efficiently than those with low-capacity (Bengson & Mangun, 2011; Burgoyne & Engle, 2020; Coleflesh & Conway, 2007; Heitz & Engle, 2007). One *may* argue that listeners require time to build the integrity of these "templates", and the switching of the target ear between blocks prevented the ability to do so for both WMC-groups (Heitz & Engle, 2007). This could be supported by Sörqvist, Nöstl & Halin (2012) results where high-capacity listeners were able to inhibit the orienting response to deviant sounds as blocks went on, but in the beginning high and low-capacity participants both experienced deviation effects to oddball sounds. While this may be the case, it is unlikely that WMC-groups would impose *no* differences at all across measures of distractibility in the presence of a *cognitive* load especially considering the controlled attention viewpoint that has been adopted here. Following the lack of WMC-group differences, it is unlikely that a *cognitive* rather than *perceptual* load was imposed during Experiment I. How then would increased distractibility under high demands be explained? It is more likely that these patterns provide evidence for the secondary explanation arguing that auditory perceptual capacities are much greater than visual.

Per the latter argument, sounds that are introduced into auditory environments will be continuously processed and/or perceived as limits of auditory perceptual capacity are more difficult to reach (Alain & Izenberg, 2003; Gomes, Barrett, Duff, Barnhardt & Ritter, 2008; Murphy, Fraenkel & Dalton, 2013; Murphy, Spence & Dalton, 2017). Increases in the complexity of sounds within auditory environments do not induce *early* methods of selective attention as is observed in the visual domain, and instead impair *late* selection abilities of listeners. Furthermore, WMCs do not appear to inform the amount of auditory perceptual processing capacity individual listeners have nor the efficiency of *late* selection in the presence of these demands as incongruent distractors were processed similarly between the groups in the absence of central demands or engagement of higher order functions. Nonetheless, WMCs do become important for overall task performance. While measured distractibility was the same across both groups, higher WMC listeners performed with better RTs and accuracy than their low-WMC counterparts indicating that even when similar intrusions from distractors are experienced, high-capacity listeners still outperform those grouped as low.

Overall, findings from Experiment I do not support the extension of perceptual load theories to the auditory domain. Perceptual demands of auditory environments do, however, negatively influence performance on complex listening tasks and should be taken into consideration when investigating task performance. Here, increased perceptual demands induced distractibility regardless of individual listener differences that would be assumed to protect or limit experienced distractibility or reduce measured distractibility.

Experiment II

Rather than assessing dichotic listening skills, the task for Experiment II was a variation of Fairnie, Moore, and Remington's (2016) auditory "search" paradigm. This paradigm served three purposes: it is similar to visual search designs from perceptual load studies, it has provided support for perceptual load in the auditory domain, and better reflects real-world listening scenarios. Load manipulations for Experiment II can be looked at in two ways: increase in the number of potential locations a target sound could be heard and/or the number of array items (Fairnie, Moore & Remington, 2016; Lavie, 2010; Murphy, Spence & Dalton, 2017). The objective of the task was to determine whether a duck's quack or a dog's bark was heard within a "frontal" array centered around zero degrees azimuth while ignoring distractions presented to the right (+90 deg) or left (-90 deg). The "frontal" array under Low-Load was only comprised of two potential locations while High-Load had four potential locations. Array locations that did not have the target stimulus for a current trial were filled with either 1kHz FM tones or white-noise bursts; therefore, under low-load the "frontal" array presented a target and one of the latter sounds while under high-load the target was surrounded by three other sounds. Like Experiment I, the first analysis was conducted to determine whether load manipulations were overall successful.

For Experiment II, RTs were longer and accuracy was poorer in the face of the four-item "frontal" array compared to the two-item array. Findings provide support that the increase in number of potential locations a sound may appear as well as number of items within an array resulted in greater demands experienced by listeners (Fairnie, Moore & Remington, 2016; Francis, 2010). Again, in order to determine the nature of these demands and delineate between *cognitive* and *perceptual* load theory predictions, differences in distractibility were evaluated; however, influence (i.e., ΔRT-Load) from the addition of background noise or the potential secondary load manipulation will be reviewed first.

Unlike Experiment I and unsupportive of hypothesis two, ΔRT-Load values did not significantly differ among Quiet, SSN nor MTB conditions. While differences were not significant, ΔRT-Load was slightly smaller in the presence of SSN and MTB compared to Quiet as can be seen in Figure 6 printed in the previous results section. This slight reduction in ΔRT-Load might suggest that background noise increased the complexity of the low-load task as described within Experiment I, but did not reach significance as load effects on performance (i.e., changes in RT) between the four-item and two-item spatial arrays were not as strong as the four-feature comparison versus two-feature identification of the dichotic listening task. Regardless, the lack of significant differences here leads to the conclusion that performance on this variation of an auditory "search" task was influenced by perceptual load manipulations independently of the type of background noise. The next factor explored was ΔRT-Distraction which allowed for finer grained speculation of the form of load (i.e., *cognitive* versus *perceptual*) imposed by the auditory "search" task.

Findings showed no differences in ΔRT-Distraction between the High-Load four-item search and the Low two-item search, indicating that influence from the incongruent distractor was the same regardless of perceptual load. The lack of ΔRT-Distraction differences contrasts both assumptions of the perceptual load theory as well as findings from Fairnie, Moore and Remington's (2016) study that motivated the development of Experiment II's task. At first glance, one may argue that the absence of ΔRT-Distraction differences resulted from an insufficient or *weak* manipulation of load (Murphy, Fraenkel & Dalton, 2013; Murphy, Spence & Dalton, 2017) as effect sizes of ΔRT-Load were small across both Accuracy and RT measurements; however, overall task performance was significantly impaired by the increase to four-item search from a two-item search. Additionally, while Fairnie, Moore and Remington's (2016) study included array set sizes ranging from one to six-items, there was still a significant difference in performance between two- and four-items providing further support that this was a successful manipulation of perceptual load. Two key differences between the present paradigm and the "original" paradigm that may have resulted in these opposing patterns of distraction are: 1) their task assessed *awareness* of a secondary sound, where distraction here was induced via response competition, and 2) "filler" array items of their task were similar to target sounds (i.e., other animal noises), while "filler" sounds for the present design were highly dissimilar (i.e., 1kHz FM tone or white-noise bursts).

Awareness reports serve as measures if inattentional deafness, where sounds beyond a central array are not heard or perceived following changes in perceptual demands of tasks (Francis, 2010; Forster & Lavie, 2009; Lavie, 2010). While these investigations are useful as they mirror a listening scenario one might encounter when watching television and missing their spouse trying to talk to them from another room, they are accompanied by a few critiques. The validity of these critiques are questionable though, primarily because of the lack of direct manipulations or number of investigations of auditory perceptual load; however, they are worth mentioning here as the present design may provide an answer to at least part of these issues. One perspective of awareness reports argues that they impose a dual-task, meaning that the sound that is being probed to determine if a listener is "deaf-to/unaware-of" often requires a response (Murphy, Fraenkel & Dalton, 2013; Murphy, Spence & Dalton, 2017). In the case of the Fairnie, Moore and Remington (2016) study, a secondary portion of the task was to determine when a "car sound" was heard. Since listeners were told to attend to this sound as well as provide a response, it seems plausible that a secondary task or a dual-task may have been imposed.

These researchers also offered within their discussion that inattentional deafness may not hold should sounds beyond an auditory search be *meaningful*. Examples provided in their work were fire alarms or car horns that are encountered to warn listeners about a change in their environment. They questioned whether the ability for auditory perceptual demands of their search task to reduce perception of the "car sound" resulted from the non-meaningful relationship between that "car sound" and items of the array or potential relevance to a listener's environment (i.e., a warning of some sort). While response competition is different in nature from awareness reports, as sounds are supposed to be ignored and not responded to, incongruent distractors do share overlap with target sounds which implies that they are particularly relevant for listeners – albeit they are not "alarming". Since distraction was observed across High- and Low- load manipulations within Experiment II, it is likely that sounds that are meaningful for listeners will induce distraction or break through a barrier that is not modulated by perceptual demands (Driver, 2001; Treisman, 1964; Spence & Santangelo, 2010). This of course cannot be concluded without some *awareness* measure, as intrusions may have been subconscious to listeners. One potential resolution without introducing response competing distractors as a secondary target may be reports of listening effort – where increased effort would lead to the conclusion that sounds were consciously perceived.

The second major difference between their design and the present task for Experiment II was that "filler" array items were similar to target sounds. This may become problematic as definitions for what serves as an auditory perceptual load manipulation is still questioned. Within their work, it is likely that the *level of similarity* paired with *number of items in a search array* imposed compounded effects making the search much more difficult across two manipulations (Lavie, 2010; Lavie, Hirst, de Fockert & Viding, 2004; Murphy, Spence & Dalton, 2017). The present study included array items that were very *dissimilar* and distractibility did not differ between the two- and four-item search arrays, so it does not appear that the addition of array items alone induces enough perceptual demand to constrain focal attention to a central search array and reduce instances of distraction. Instead, similarity or a secondary load manipulation is required to observe patterns from visual or cross-modal perceptual load designs. Perhaps a solution to this would be to keep the number of array items the same and vary the similarity between the items included (i.e., low-load = tones, high-load = environmental sounds). This also appears to corroborate the argument that auditory perceptual capacities are not as limited as visual. Again, a benefit of the current work is that measures of WMCs were included to assist the interpretation of findings, types of load, as well as realistic implications for listeners.

Analysis of WMCs yielded significant effects of WMC-group across ΔRT-Load, ΔRT-Distraction, and overall task performance. ΔRT-Load's were greater for low-WMC listeners and neared zero for those grouped as high-WMC, meaning that low-WMC listeners were the only group affected by the increase from the two-item array to the four-item. Interestingly, ΔRT-Distraction values were greater for high-WMC listeners and now neared zero for low-WMC listeners. This result was surprising as measured distractibility is typically reduced for high- compared to low-WMC groups (Heitz & Engle, 2007; Sörqvist, 2010; Sörqvist, Nöstl & Halin, 2012), while here low-WMC listeners did not experience intrusions from incongruent or response competing distractors and high-WMC listeners did. This increased distraction for the high-capacity group as well as reduced distractibility for the low-capacity group was not modulated by perceptual load - meaning low-capacity listeners never experienced distraction and high-capacity listeners always experienced distraction. Another interesting finding was that even though high-capacity listeners expressed significant ΔRT-Distraction values, their overall task performance was faster (lower RTs) and more accurate than their low-capacity counterparts. These group effects of Experiment II elicit some interesting discussion points. Perhaps the perceptual load theory *does* transfer to the auditory domain but effects are dependent on individual differences that are often left out of these conversations. From another perspective, these differences may shed light on processing or strategies that listeners with varying central limitations employ to achieve optimal task performance.

First, it very well may be the case that predictions of perceptual load replicated within Experiment II. Initial analysis of ΔRT-Distraction showed no effects of load (i.e., ΔRT-Distraction measures were the same whether load was high or low), as capacity group differences were major contributors and effects averaged out when these groups were collapsed. This would suggest that WMCs are also informative of the amount of perceptual capacity listeners have at least during auditory search. Following this potential interpretation, low-WMC listeners perceptual capacities were reached by the bottom-up demands of the task which constrained their attention to the "frontal" search array and inhibited perception of distracting items. This was not observed for high-WMC listeners as their perceptual capacities were not exhausted and additional perception beyond the "frontal" array was allowed to continue. Here, *early* selection would have occurred for low-capacity listeners, where *late* selection would have occurred for high-capacity listeners. While this explanation intuitively makes sense and pairs with the results well, there are a few conceptual issues that caution this from being fully adopted.

Per traditional load theory accounts increased distractibility would be expected to increase when perceptual load was low. In the case of low-WMC listeners, there was a significant effect of load meaning that demands of the two-item search were less than those imposed by the four-item search. However, effects of distraction did not differ across load manipulations for the low-WMC group. This suggests that perceptual load was not observed – as distractibility would have been seen within the two-item search. One may argue that the two-item search imposed a high perceptual load from the start, where low-WMC listeners capacities were taxed even under low-load and only increased with the four-item array. This would explain the lack of changes in distractibility, as well as the reduced effects of distraction had perceptual loads been high across both arrays. This is unlikely as effects were significant for overall performance via RTs and accuracy and a two-item search really should not be all that perceptually demanding according to Fairnie, Moore, and Remington (2016) results that presented two-item searches as relatively *easy*. Additionally, had low-WMC informed auditory perceptual capacity limitations it is likely that some WMC-group differences would have also been observed within Experiment I, which in fact was not the case. Perhaps exploring contributions from WMC-groups along a spectrum of load manipulation within an auditory search array like to the one imposed by Fairnie, Moore and Remington (2016) would help with this.

Following these interpretations and factors, I am more inclined to argue that these WMC group differences resulted from different strategies employed by the WMC groups to perform the task. Along this thought path, *reduced* distractibility for low-WMC listeners would have resulted from the engagement of attentional control where high-WMC did not have to deploy higher-order attentional mechanisms to be just as successful on the listening task of Experiment II irrespective of how much distractibility they experienced.

Overall, findings from Experiment II do not align with traditional visual perceptual load predictions as distractibility did not differ between High- and Low- load conditions. Additionally, the inclusion of background noise had no effects on performance across either ΔRT-Load or ΔRT-Distraction indicating that negative effects are limited to tasks with central components. Here, it appears that WMCs were more indicative of strategies employed during simple low-level auditory search tasks rather than influence from perceptual demands alone; where low-WMC listeners required active attentional control and high-WMC listeners did not.

General Discussion

Despite the differences in results between Experiments I and II, patterns were largely unsupportive of the extension of the perceptual load theory to the auditory domain. Rather than constraining focal attention, or inducing *early* selection, increased bottom-up demands of the auditory tasks did not modulate methods of selection as auditory selective attention operated *late* in the presence of isolated perceptual loads. In other words, the "gatekeeper" selective attention automatically granted access to higher levels of processing for auditory inputs and will continuously do so regardless of their complexities (Experiment I) and/or number of sounds within an environment (Experiment II). As more auditory information becomes introduced to perceptual processing systems, the efficiency of *late* selection is reduced which resulted in increased distractibility in the presence of high perceptual loads. This is not to say that *early* methods of auditory selection never take place. The argument here is that this method is not informed by bottom-up perceptual demands, and instead requires some form of top-down engagement in order to be observed in the auditory domain.

A recurring statement throughout this narrative has been that the goal of *perceptually* loaded tasks are to limit engagement from higher order functions to directly investigate contributions from bottom-up stimulus features alone (Lavie, 2010; Lavie, Hirst, de Fockert & Viding, 2004; Lavie & Tsal, 1994). The biggest question with respect to audition centers around the unknown definitions of what constitutes perceptual load manipulations of sounds as there is a fine line between *cognitive* and *perceptual* demands (Francis, 2010; Murphy, Fraenkel & Dalton, 2013; Murphy, Spence & Dalton, 2017). General conclusions from studies yielding similar results to the current work, or that provide evidence against the applicability of the perceptual load theory to the auditory domain, have been challenged along these lines (Francis, 2010; Gomes, Barrett, Duff, Barnhardt & Ritter, 2008; Murphy, Fraenkel & Dalton, 2013) – especially those that implemented tasks similar to Experiment I where multiple features of sounds must be compared. This has been partially resolved by the current study as measures of WMCs were included to assist the ability to define the loads imposed across both tasks, as WMCs are expected to be sensitive to the presence of cognitive demands as well as expose whether top-down functions were engaged.

Beginning with Experiment I, the lack of influence from WMC-groups on both ΔRT-Load and ΔRT-Distraction measures support the notion that task demands were *perceptual* in nature. During the development of Experiment I's dichotic listening task, the ability for listeners to recruit top-down functions was controlled for by switching the to-be-attended-to target ear between each block. When auditory inputs are only presented to one ear for the entire duration of a task, listeners can employ attentional control mechanisms to "zero-in on" or constrain focal attention towards that target direction (Bookbinder & Osman, 1979; Coleflesh & Conway, 2007; Heitz & Engle, 2007; Sörqvist, 2010); however, this "build-up" requires time which was essentially removed for Experiment I. With the ability to develop a "build-up" removed, perceptual demands of the stimuli presented in the target ear became the primary factor that could constrain focal attention. In the absence of top-down engagement, perceptual demands alone were unable to constrain focal attention. Additionally, arguments suggesting that *cognitive* rather than *perceptual* loads are imposed by increases in the number of stimulus features to be compared (i.e., *perceptual operations required by tasks*) do not align with the present results. As established in the prior discussion, had *cognitive* rather than *perceptual* demands been imposed by the need to develop and retain goal-directed "templates", the high-WMC listener group would have shown reduced effects of load and distractibility (i.e., smaller ΔRT-Load and ΔRT-Distraction values) than their low-WMC counterparts (Engle, 2018). This provides support for the present argument that increases in complexities of sounds, or auditory perceptual loads, only add to the inefficiency of *late* selection and can become detrimental for listeners – which opposes visual perceptual load findings.

Patterns of Experiment II were slightly different and less robust following an effect size analysis than those observed in Experiment I, but arrived at the same conclusion in that traditional perceptual load theories did not transfer to the auditory domain. Rather than ΔRT-Distraction values being larger under High- than Low-perceptual conditions seen in Experiment I, they did not differ as a factor of load manipulations for Experiment II. The argument that *perceptual* demands were induced within Experiment II by changing the number of locations a sound may appear in a "frontal" array or increasing the number of items in an auditory search are not as clear cut as Experiment I, but were assumed following evidence provided by similar work (Fairnie, Moore & Remington, 2016; Murphy, Fraenkel & Dalton, 2013). The most interesting finding from Experiment II was that *perceptual* demands did not inform performance, instead individuals WMCs contributed to measures of distractibility. Within each listener group, ΔRT-Distraction was the same during the two- and four-item "frontal" arrays. Contrary to typical WMC effects, distractibility was *reduced* for those categorized with low-WMC and *increased* for high-WMC listeners. Additionally, ΔRT-Load measures were significantly greater for low-WMC listeners and non-existent for the high-WMC group. Significant ΔRT-Load values counter the potential argument that low-WMC listeners were operating at perceptual limits from the start even during the two-item search, as this should have resulted in at least some ΔRT-Distraction differences. The control limiting top-down engagement placed within Experiment I could not be applied for Experiment II, allowing for listeners to recruit higher-order mechanisms should they be required. Here, the difficulty of the search task as a whole - regardless of perceptual load - resulted in the engagement of top-down attentional control from low-WMC listeners, but not high-WMC listeners as they were still able to efficiently perform the task without the need to "kick-in" attentional control. This difference in strategy deployment is further supported by results showing overall faster RTs and better accuracy for high-WMC listeners even though they experienced more distractibility. Therefore, active top-down control exerted by the low-WMC group likely reduced their perception of distractor sounds instead of bottom-up demands.

Thus far, evidence is in strong favor against the applicability of the perceptual load theory to the auditory domain. Instead, task demands accompanied by requirements of top-down recruitment are more informative of instances of distractibility than bottom-up complexities. This is highlighted by the differences in the underlying nature of the listening tasks employed within Experiments I and II. The dichotic listening task of Experiment I limited the need for attentional search as well as prevented the "build-up" of the integrity of focal attention. Attentional search was restricted during dichotic listening as instructions served as a *cue* directing attention towards a target ear, and "build-up" was prevented by changing the target ear between each block. This inability to recruit top-down functions resulted in increased distractibility for both high- and low-WMC listeners in the presence of high perceptual demands. Similarly, the argument from Experiment II states that in an undirected listening search, high-WMC listeners could perform optimally without the need to exert control. Again, in the absence of control high-WMC listeners experienced more distractibility; however, low-WMC listeners *did* need to exert control to complete the auditory search regardless of the load of the "frontal" array thereby reducing distractibility. While these findings do not align with perceptual load theory predictions, they do agree with the general assumptions of Task-Engagement-Distractor Trade-Off (TEDTOFF) (Sörqvist & Marsh, 2015; Sörqvist &

Rönnberg, 2014) model and provide new insights to controlled attention views of WMCs as well as potential implications for listeners.

Findings align with the TEDTOFF model, as the only instance of reduced distractor effects were observed when higher-level or active top-down task engagement was assumed. According to the TEDTOFF model, the ability to mediate distractors is directly related to levels of concentration participants reach during completion of tasks – where increased concentration reduces distractibility (Halin, Marsh, Hellman, Hellström & Sörqvist, 2014; Sörqvist & Marsh, 2015; Sörqvist & Rönnberg, 2014). Concentration, or magnitude of task-engagement, is typically elicited by modulating memory loads within tasks such as n-backs or the inclusion of a coordination component such as a dual-task; however, Sörqvist and Rönnberg (2014) also stated that visual perceptual load findings fit within their model as central perceptual demands also induced higher concentration. What is interesting from the current work, is that solely auditory perceptual demands do not induce this same task engagement – especially when the ability to deploy top-down mechanisms is removed (i.e., Experiment I). Additionally, both the TEDTOFF and controlled attention views of WMCs always assume that higher-WMC individuals will exhibit reduced distractor interference as they are said to have the ability to reach higher levels of task engagement as well as maintain this engagement for longer periods of time. This makes sense, as the tasks individuals are typically presented with in cognitive hearing science work encourage/require top-down resources to be successful and high-WMC individuals have more flexible control over these processes.

Findings from the current work, specifically from Experiment II, appear to add to what this "flexible control" means for high-WMC listeners.

Across both Experiments I and II, high-WMC listeners either experienced the same amount or *more* interference from response competing distractors than the low-WMC listening group, and yet high-WMC listeners were still able to perform *faster* and more *accurately* overall on both listening tasks. While the ability to engage top-down functions was limited by Experiment I, it was freely allowed within Experiment II. Following the idea of "flexible control", high-WMC listeners were able to judge or assess the task of Experiment II and did not have to engage top-down control to perform well, as their perception of distractors did not impair overall task performance. In contrast, low-WMC with less flexibility had to engage control to efficiently search. Instead of solely defining high-WMC as reflecting superior abilities to engage active top-down interventions, it appears that this measure also reflects the ability to restrain the deployment of additional executive functions that are not required by task demands. Perhaps this ability to restrain top-down deployment results in the ability to maintain task engagement for longer periods of time as well as reduced reports of listening effort during complex listening tasks. While this concept is interesting and warrants further investigation as it can potentially add to controlled attention views, it remains underdeveloped as this was not the primary objective of the present study which was to determine whether predictions of the perceptual load theory transfer to the auditory domain.

Overall, predictions of perceptual load theories did not transfer to the auditory domain. Loads here are most likely *perceptual* in nature as WMC-group effects were absent in Experiment I and were unrelated to anticipated perceptual effects in Experiment II. An additional factor of interest was if the addition of background noise changed the type of load imposed by a listening task. For both Experiments I and II, neither MTB nor SSN influenced task performance through measures of distractibility. Had background noise changed the nature of the task load from *perceptual* to *cognitive*, measures of distractibility would have differed. The lack of influence from background noise likely resulted from the absence of a central *cognitive* demand of the task, as most background noise influence during complex listening tasks has this component (i.e., speech recognition, attention switching, etc.).

General conclusions of the current work support the notion that perception of bottom-up auditory inputs will persist regardless of perceptual load in the absence of top-down engagement. Studies that provide support of the traditional perceptual load theory in the auditory domain instead appear to have a component that elicits top-down control or engagement. For example, the *shadowing* or repetition of speech during the "cocktail party effect" likely recruited higher-order processing that allowed for the rejection of information presented to the non-attended (i.e., to-be-ignored) ear (Cherry, 1953; Cowan et al, 2005; Driver, 2001; Moray, 1959; Spence & Santangelo, 2010). Here, the inclusion of speech is what would have required higher-order processing. Also considering Fairnie, Moore and Remington's (2016) study that appears to provide a steadfast account of this theory's extension, top-down engagement may have been required by the shape of their auditory search as it extended beyond a "frontal" location into a semi-circle around the listener, inclusion of up to six locations, as well as the addition of array items that were from the same category as targets (i.e., all were animal sounds). These traits of their study likely added to the ambiguity of the search for a target sound and recruited attentional control, leading to increased *engagement* or *concentration* and reduced perception of their "critical" sound.

Rather than resulting in uniform patterns of distraction consistent with anticipated load theory predictions, results from Experiments I and II suggest that various perceptual manipulations and/or tasks impose dissociable effects that are situationally dependent. The latter point, paired with arguments from related work discussed above, makes it difficult to confine auditory perceptual influence into a unitary theoretical framework unlike visual. Despite the inability to hone in on a singular theory of perceptual load specific to audition, presumably due to the argument that our auditory systems are constantly surveying our environments (Murphy, Fraenkel & Dalton, 2013; Murphy, Spence & Dalton, 2017), it is important to note that auditory *perceptual* demands did influence performance differently than *cognitive*. This became apparent when evaluating novel contributions from WMCs, which provided evidence within the current study indicating that different mechanisms are recruited to manage these two categories of load. Additional investigations are necessary to aid the generalizability of these claims and work towards the development of a holistic definition of perceptual load that may be applicable to the realm of audition.

Chapter 6: Conclusion

Interactions with complex listening environments are unavoidable, and listeners express variable reports of difficulty and/or success. Considering aging and hearing-impaired populations, these scenarios are often accompanied by frustration which unfortunately can lead to withdrawal from daily interactions and more isolated lifestyles. The objectives of audiologists, hearing health care professionals, and cognitive hearing scientists are to develop aural rehabilitation programs as well as tools that can be used to alleviate listener complaints and promote better quality of life among clinical populations (Arlinger, Lunner, Lyxell & Pichora-Fuller, 2009; Peelle, 2018; Pichora-Fuller et al., 2016; Rönnberg et al., 2013; Rönnberg, Holmer & Rudner, 2019; Shinn-Cunningham & Best, 2008; Strauss & Francis, 2017). This book was developed with this patient-centered motive in mind, albeit not directly investigated, and approached these issues through the lens of perceptual load, selective attention, and auditory distraction.

Perceptual load investigations have two purposes with respect to audition. First, real-world auditory schemes are comprised of both *cognitive* and *perceptual* demands; however, separable effects of these demands have yet to be established in a definitive manner. Second, traditional perceptual load theory accounts assume *reduced* distractibility when demands are *high* – this reduction is driven by bottom-up complexities independent of recruitment of top-down functions and is described as *passive* or *effortless.* In the pursuit of identifying factors that contribute to *easier* listening, perceptual load was manipulated directly across Experiments I and II.

Findings from both experiments were unsupportive of the transferability of the perceptual load theory to the auditory domain. Instead, increased auditory perceptual load impaired the efficiency of selective attention resulting in greater intrusions from distracting sounds. While increased distractibility is also observed under *cognitive* loads, results from the current work provide evidence that dissociates how listeners interact with *cognitive* versus *perceptual* demands. Specifically, WMCs typically relate to distractor mediation in the face of *cognitive* loads; the absence of WMC-group differences here highlights differences in both processing of bottom-up *perceptual* demands as well as underlying mechanisms that contribute to listener performance. Additionally, the addition of background noise conditions did not influence performance on these perceptual tasks in the same way that has been demonstrated under cognitive demands. Therefore, future work within this realm should carefully consider how much of a task is influenced by *perceptual* and *cognitive* demands before drawing firm conclusions regarding listener performance. For example, what would outcomes look like should *perceptual* features outweigh *cognitive*? Perhaps this unequal distribution is driven by internal listener traits – where listeners are presented with the same *task,* but individual differences drive which type of demand is most influential over their performance. This was partially observed within Experiment II, where WMC-groups appeared to inform different strategies employed during simple auditory search even for young normal hearing listeners – it is even more likely that these factors would play important roles for clinical populations.

Apart from the prescription of amplification devices, a common approach to improving listener experiences is the development of training paradigms and/or programs. A major issue is the inability to achieve "far-transfer" or generalizability of these tasks to mechanisms separate from the one engaged during training. Emerging findings from the present work encourages consideration of individual differences as well as greater exploration into dissociative demands of real-world listening tasks, as "one-size-fits-all" approaches seem to be ineffective. This is one possible direction that can be moved towards following additional auditory perceptual load studies similar to this book to further clarify what defines "auditory perceptual load" and that include other individual difference measures (i.e., single WMC scores versus collapsed groups), various age groups, and degrees of hearing-impairment.

References

Alain, C. & Izenberg, A. (2003). Effects of attentional load on auditory scene analysis. *Journal of Cognitive Neuroscience, 15*(7), 1063-1073. doi: 10.1162/08989290377 0007443

Anderson, M. C. (2001). Active forgetting: Evidence for functional inhibition as a source of memory failure. *Journal of Aggression, Maltreatment, & Trauma, 4*, 185-210. doi: 10.1300/J146v04n02_09

Arlinger, S. Lunner, T. Lyxell, B. & Pichora-Fuller, M. K. (2009). The emergence of Cognitive Hearing Science. *Scandinavian Journal of Psychology, 50*, 371-284. doi: 10.111/j.1467-9450.2009.00753.x

Atcherson, S. R., Nagaraj, N. K., Kennett, S. E. & Levisee, M. (2015). Overview of central auditory processing deficits in older adults. *Seminars in Hearing, 36*(3), 150-161. doi: 10.1055/s-0035-1555118

Awh, E., Vogel, E. K. & Oh, S. H. (2006). Interactions between attention and working memory. *Neuroscience, 139*(1), 201-208. doi: 10.1016/j.neuroscience.2005.08. 023

Baddeley, A. (2010). Working memory. *Current Biology, 4*(23), 136-140. doi: 10.1016/ j.cub.2008.12.014

Bengson, J. J. & Mangun, G. R. (2011). Individual working memory capacity is uniquely correlated with feature-based attention when combined with spatial attention. *Attention, Perception, & Psychophysics, 73*(1), 86-102. doi: 10.3758/s13414-010- 0020-7

Bookbinder, J. & Osman, E. (1979). Attentional strategies in dichotic listening. *Memory & Cognition, 7,* 511-520.

Bowers, J. S. & Davis, C. J. (2004). Is speech perception modular or interactive? *Trends in Cognitive Sciences, 8*(1), 3-5. doi: 10.1016/j.tics.2003.10.018

Broadbent, D. E. (1958). *Perception and Communication.* Oxford: Oxford University Press.

Bronkhorst, A. W. (2015). The cocktail-party problem revisited: early processing and selection of multi-talker speech. *Attention, Perception and Psychophysics, 77,* 1465-1487. doi: 10.3758/s13414-015-0882-9

Burgoyne, A. P. & Engle, R. W. (2020). Attention control: A cornerstone of higher-order cognition. *Current Directions in Psychological Science, 29*(6), 624-630. doi: 10.177/0963721420969371

Callahan, C. M., Unverzagt, F. W., Hui, S. L., Perkins, A. J. & Hendrie, H. C. (2002). Six-item screener to identify cognitive impairment among potential subjects for clinical research. *Medical Care, 40*(9), 771-781. doi: 10.1097/00005650-200209000-00007

Carlile, S., Fox, A., Orchard-Mills, Leung, J. & Alais, D. (2016). Six degrees of auditory spatial separation. *Journal of the Association for Research in Otolaryngology, 17*(3), 209-221. doi: 10.1007/s10162-016-0560-1

Cherry, E. C. (1953). Some experiments in the recognition of speech with one and two ears. *Journal of the Acoustical Society of America, 25,* 975-979.

Coleflesh, G. J. H. & Conway, A. R. A. (2007). Individual differences in working memory capacity and divided attention in dichotic listening. *Psychonomic Bulletin & Review, 14*, 699-703.

Conway, A. R. A., Cowan, N. & Bunting, M. F. (2001). The cocktail party phenomenon revisited: The importance of working memory capacity. *Psychonomic Bulletin & Review, 8*(2), 331-335.

Cowan, N. (1999). *An Embedded-Process Model of working memory*. In A. Miyake & P. Shah (Eds.), *Models of working memory: Mechanisms of active maintenance and executive control* (p. 62-101). Cambridge University Press.

Cowan, N. (2010). The magical mystery four: How is working memory capacity limited, and why? *Current Directions in Psychological Sciences, 19*(1), 51-57. doi: 10.1177/0963721409359277

Cowan, N. (2017). The many faces of working memory and short-term storage. *Psychonomic Bulletin and Review, 24*(4), 1158-1170. doi: 10.3758/s13429-019-1191-6

Cowan, N., Elliott, E. M., Saults, J. S., Morey, C. C., Mattox, S., Hismjatullina, A. & Conway, A. R. A. (2005). On the capacity of attention: Its estimation and its role in working memory and cognitive aptitudes. *Cognitive Psychology, 51*(1), 42-100.

Craik, F. I. M. (2014). Effects of distraction on memory and cognition: a commentary. *Frontiers in Psychology, 5*(841), 1-5. doi: 10.1016/j.cogpsych.2004.12.001

Dai, L. Best, V. & Shinn-Cunningham, B. G. (2018). Sensorineural hearing loss degrades behavioral and physiologic measures of human spatial selective auditory attention. *Proceedings of the National Academy of Sciences of the United States of America, 115*(14), 3286-3295. doi: 10.1073/pnas.1721226115

Dalton, P., Santangelo, V. & Spence, C. (2009). The role of working memory in auditory selective attention. *Journal of Experimental Psychology, 62*(11), 2126-2132. doi: 10.1080/17470210903023646

Daneman, M. & Carpenter, P. A. (1980). Individual differences in working memory and reading. *Journal of Verbal Learning and Verbal Behavior, 19*(4), 450-466.

Deustch, J. A. & Deustch, D. (1963). Attention: Some theoretical considerations. *Psychological Review, 70*, 80-90.

DiGiovanni, J. J., Riffle, T. L., Lynch, E. E. & Nagaraj, N. K. (2017). Noise characteristics and their impact on working memory and listening comprehension performance. *Proceedings of Meetings of Acoustics, 30*(1), 1-14. doi: 10.1121/2.0000597

DiGiovanni, J. J., Riffle, T. L., Weaver, A. J. & Lynch, E. E. (2017). Relating verbal and non-verbal auditory spans to language comprehension performance. *Proceedings of Meetings of Acoustics, 30*(1), 1-11. doi: 10.1121/2.0000628

Dong, S., Reder, L. M., Yau, Y., Liu, Y. & Chen, F. (2015). Individual differences in working memory capacity are reflected in different ERP and EEG patterns to task difficulty. *Brain Research, 7*(1616), 146-156. doi: 10.1016/j.brainres.2015.05.003

Dos Santos Sequeira, S., Specht, K., Hämäläinen, H. & Hugdahl, K. (2008). The effects

of background noise on dichotic listening to consonant-vowel syllables. *Brain and*

Language, 107(1), 11-15. doi: 10.1016/j.bandl.2008.06.001

Driver, J. (2001). A selective review of selective attention research from the past century.

British Journal of Psychology, 92(1), 53-78. doi: 10.1348/000712601162103

Engle, R. W. (2018). Working memory and executive attention: A revisit. *Perspectives*

on Psychological Science, 13(2), 190-193. doi: 10.1177/1745691617720478

Erikson, B. A. & Erikson, C. W. (1974). Effects of noise letters upon identification of a

target letter in a nonsearch task. *Perception and Psychophysics, 16*(1), 143-149.

Erikson, M. A., Ruffle, A. & Gold, J. M. (2016). Meta-analysis of mismatch negativity

in schizophrenia: from clinical risk to disease specificity and progression.

Biological Psychiatry, 79(12), 980-987. doi: 10.1016/j.biopsych.2015.08.025

Fairnie, J., Moore, B. C. J. & Remington, A. (2016). Missing a trick: Auditory load

modulates conscious awareness in audition. *Journal of Experimental Psychology:*

Human Perception and Performance, 42(7), 930-938. doi: 10.1037/xhp0000204

Francis, A. L. (2010). Improved segregation of simultaneous talkers differentially affects

perceptual and cognitive capacity demands for recognizing speech in competing

speech. *Attention, Perception & Psychophysics, 72*(2), 501-516. doi:

10.3758/APP.72.2.501

Francis, A. L., Bent, T., Schumaker, J., Love, J. & Silbert, N. (2021). Listener characteristics differentially affect self-reported and physiological measures of effort associated with two challenging listening conditions. *Attention, Perception and Psychophysics,* 1-24. doi: 10.3758/s13414-020-02195-9

Francis, A. L. & Love, J. (2019). Listening effort: Are we measuring cognition, affect, or both? *WIREs Cognitive Science,* 1-27. doi: 10.1002/wsc.1514

Forster, S. & Lavie, N. (2009). Harnessing the wandering mind: The role of perceptual load. *Cognition, 111*(3), 345-355. doi: 10.1016/j.cognition.2009.02.006

Fougine, D. (2008). The relationship between attention and working memory. *New Research in Short-Term Memory, 1-45.*

Garofalo, S., Battaglia, S. & di Pellegrino, G. (2019). Individual differences in working memory capacity and cue-guided behavior in humans. *Scientific Reports, 9*(7327), 1-14. doi: 10.1038/s41598-019-43860-w

Gomes, H., Barrett, S., Duff, M., Barnhardt, J. & Ritter, W. (2008). The effects of interstimulus interval on event-related indices of attention: An auditory selective attention test of the perceptual load theory. *Clinical Neurophysiology, 119*(3), 542-555. doi: 10.1016/j.clinph.2007.11.014

Guijo, L. M., Horiuti, M. B., Nardez, T. M. B. & Cardoso, A. C. V. (2018). Listening effort and working memory capacity in hearing impaired individuals: An integrative literature review. *Speech, Language, Hearing Sciences and Education Journal, 20*(6), 798-807. doi: 10.1590/1982-021620182066618

107

Halin, N. Marsh, J. E., Hellman, A. Hellström, I. & Sörqvist, P. (2014). A shield against distraction. *Journal of Applied Research in Memory and Cognition, 3*(1), 31-36. doi:10.1016/j.jarmac.2014.01.003

Hasher, L., Lustig, C. & Zacks, R. (2007). *Inhibitory mechanisms and the control of attention.* In A. R. A. Conway, C. Jarrold, M. J. Kane (Eds.) & A. Miyake & J. N. Towse (Ed.), *Variation in working memory* (p. 227-249). Oxford University Press.

Heitz, R. P. & Engle, R. W. (2007). Focusing the spotlight: Individual differences in visual attention control. *Journal of Experimental Psychology: General, 136*(2), 217-240.

Herwig, N. A. & Bunzeck, N. (2015). Differential effects of white noise in cognitive and perceptual tasks. *Frontiers in Psychology, 3*(6), 1-16. doi: 10.3389/fpsyg.2015. 01639

Hughes, R. W. (2014). Auditory distraction: A duplex-mechanism account. *PsyCh Journal, 3*(1), 30-41. doi: 10.1002/pchj.44

Hughes, R. W., Hurlstone, M. J., Marsh, J. E., Vachon, F. & Jones, D. M. (2013). Cognitive control of auditory distraction: Impact of task difficulty, foreknowledge, and working memory capacity supports duplex-mechanism account. *Journal of Experimental Psychology: Human Perception and Performance, 39*(2), 539-553. doi: 10.1037/a0029064

Humes, L. E., Kidd, G. R. & Lentz, J. J. (2013). Auditory and cognitive factors underlying individual differences in aided speech-understanding among older adults. *Frontiers in Systems Neuroscience, 7*(55), 1-16. doi: 10.3389/fnsys. 2013.00055

Kane, M. J., Bleckley, M. K., Conway, A. R. & Engle, R. W. (2001). A controlled-attention view of working memory capacity. *Journal of Experimental Psychology, 130*(2), 169-183. doi: 10.1037//0096-3445.130.169

Kane, M. J. & Engle, R. W. (2003). Working-memory capacity and the control of attention: the contributions of goal neglect, response competition, and task set to Stroop interference. *Journal of Experimental Psychology, 132*(1), 47-70. doi: 10.1037/0096-3445.132.1.47

Konstantinou, N., Beal, E., King, J. R. & Lavie, N. (2014). Working memory load and distraction: dissociable effects of visual maintenance and cognitive control. *Attention, Perception & Psychophysics, 76*, 1985-1997. doi: 10.3758/s13414-014-0742-z

Lavie, N. (2010). Attention, distraction and cognitive control under load. *Current Directions in Psychological Science, 19*, 143-148. doi: 10.1177/096372 1410370295

Lavie, N., Beck, D. M. & Konstantinou, N. (2014). Blinded by the load: Attention, awareness and the role of perceptual load. *Philosophical Transactions of the Royal Society B Biological Sciences, 369*(1641), 1-11. doi: 10.1098/rstb.2013.0205

Lavie, N., Hirst, A., de Fockert, J. W. & Viding, E. (2004). Load theory of selective attention and cognitive control. *Journal of Experimental Psychology: General, 133*, 339-354. doi: 10.1037/0096-3445.133.3.339

Lavie, N. & Tsal, Y. (1994). Perceptual load as a major determinant of the locus of selection in visual attention. *Perception & Psychophysics, 56*, 183-197.

Le Prell, C. G. & Clavier, O. H. (2017). Effects of noise on speech recognition: Challenges for communication by service members. *Hearing Research, 349*, 76-89. doi: 10.1016/j.heares.2016.10.004

Lemke, U. & Besser, J. (2016). Cognitive load and listening effort: Concepts and age-related considerations. *Ear and Hearing, 37*, 77-84. doi: 10.1097/AUD.00000000000000304

Lewis, J. D., Kopun, J., Neely, S. T., Schmid, K. K. & Gorga, M. P. (2015). Tone-burst auditory brainstem response wave V latencies in normal-hearing and hearing-impaired ears. *Journal of the Acoustical Society of America, 138*(5), 3210-3219.

Li, Z. & Lou, J. (2019). Flanker tasks based on congruency manipulation are biased measures of selective attention in perceptual load studies. *Attention, Perception & Psychophysics, 81*, 1836-1845. doi: 10.103758/s13414-01730-7

Lunner, T. (2003). Cognitive function in relation to hearing aid use. *International Journal of Audiology, 42*, 49-58. doi: 10.3109/14992020309074624

Lynch, E. E. & DiGiovanni, (2019, March). Effects of perceptual load and task expectations on distractor interference. Abstract. American Auditory Society 45, 115.

Macdonald, J. S. P. & Lavie, N. (2011). Visual perceptual load induces inattentional

deafness. *Attention, Perception, and Psychophysics, 73*, 1780-1789. doi:

10.3758/s13414-011-0144-4

Macken, W. J., Phelps, F. G. & Jones, D. M. (2009). What causes auditory distraction?.

Psychonomic Bulletin & Review, 16(1), 139-144. doi: 10.3758/PBR.16.1.139.

Marois, A., Marsh, J. E. & Vachon, F. (2019). Is auditory distraction by changing-state

and deviant sounds underpinned by the same mechanism? Evidence from

pupillometry. *Biological Psychology, 141*, 64-74. doi: 10.1016/j.biopsycho.

2019.01.001

Mathy, F., Chekaf, M. & Cowan, N. (2018). Simple and complex working memory tasks

allow similar benefits from information compression. *Journal of Cognition, 1*(1),

1-12. doi: 10.53334/joc.31

McElree, B. (2001). Working memory and focal attention. *Journal of Experimental

Psychology: Leaning, Memory and Cognition, 27*(3), 817-835. doi: 10.1037/0278-

7393.3.817

Meier, M. E. & Kane, M. J. (2013). Working memory capacity and Stroop interference:

Global versus local indices of executive control. *Journal of Experimental

Psychology: Learning, Memory and Cognition, 39*(3), 748-759. doi:

10.1037/a0029200

Meister, H., Rählmann, S. & Walger, M. (2018). Low background noise increases

cognitive load in older adults listening to competing speech. *Journal of the

Acoustical Society of America, 144*(5), 417-422. doi: 10.1121/1.5078953

Miller, G. (1956). The magical number seven, plus or minus two: Some limits on our capacity for processing information. *The Psychological Review, 63*, 81-97.

Moray, N. (1959). Attention in dichotic listening: Affective cues and the influence of intrusions. *Quarterly Journal of Experimental Psychology, 11*, 56-60.

Murphy, S., Fraenkel, N. & Dalton, P. (2013). Perceptual load does not modulate auditory distractor processing. *Cognition, 129*(2), 345-355. doi: 10.1016/ j.cognition.2013.07.014

Murphy, G. & Greene, C. M. (2017). The elephant in the road: Auditory perceptual load affects driver perception and awareness. *Applied Cognitive Psychology, 31*, 258- 263. doi: 10.1002/acp.3311

Murphy, G., Groeger, J. A. & Greene, C. M. (2016). Twenty years of load theory – Where are we now, and where should we go next? *Psychonomic Bulletin & Review, 23*, 1316-1340. doi: 10.375/s134-015-0982-5

Murphy, S., Spence, C. & Dalton, P. (2017). Auditory perceptual load: A review. *Hearing Research, 352*, 40-48. doi: 10.1016/j.heares.2017.02.005

Myachykov, A., Scheepers, C. & Shtyrov, Y. Y. (2013). Interfaces between language and cognition. *Frontiers in Psychology, 4*, 1-2. doi: 10.3389/fpsyg.2013.00258

Näätänen, R., Sussman, E. S., Salisbury, D. & Shafer, V. L. (2014). Mismatch negativity (MMN) as an index of cognitive dysfunction. *Brain Topography, 27*(4), 451-466. doi: 10.1007/s10548-014-0374-6

Newport, E. L. (2011). The modularity issue in language acquisition: A rapprochement?

Comments on Gallistel and Chomsky. *Language, Learning, and Development,*

7(4), 279-286. doi: 10.1080/15475441.2011.605309

Oberauer, K. (2019). Working memory and attention – A conceptual analysis and review.

Journal of Cognition, 2(1), 1-23. doi: 10.5334/joc.58.

Oberfeld, D. & Klöckner-Nowotny, F. (2016). Individual differences in selective

attention predict speech identification at a cocktail party. *Elife, 5*(16747), 1-24.

doi: 10.7554/eLife.16747

Parasuraman, R. (1980). Effects of information processing demands on slow negative

shift latencies and N100 amplitude in selective and divided attention. *Biological

Psychology, 11*, 217-233.

Peelle, J. E. (2018). Listening effort: How the cognitive consequences of acoustic

challenge are reflected in brain and behavior. *Ear and Hearing, 39*(2), 204-214.

doi: 10.1097/AUD.0000000000000494

Pichora-Fuller, K. M., Kramer, S. E., Eckert, M. A., Edwards, B., Hornsby, B. W. Y.,

Humes, L. E., … Wingfield, A. (2016). Hearing impairment and cognitive energy:

The Framework for Understanding Effortful Listening (FUEL). *Ear and Hearing,

37*, 1:5S-27S. doi: 10.1097/AUD. 0000000000000312

Pollmann, S., Maertens, M., von Cramon, D. Y., Lepsien, J. & Hugdahl, K. (2002).

Dichotic listening in patients with splenial and nonsplenial callosal lesions.

Neuropsychology, 16(1), 56-64. doi: 10.1037/0894-4105.16.1.56

Repovs, G. & Baddeley, A. (2006). The multi-component model of working memory: explorations in experimental cognitive psychology. *Neuroscience, 139*(1), 5-21. doi: 10.1016/j.neuroscience.2005.12.061

Rönnberg, J., Holmer, E. & Rudner, M. (2019). Cognitive hearing science and ease of language understanding. *International Journal of Audiology, 58*(5), 247-261. doi: 10.1080/14992027.2018.1551631

Rönnberg, J., Lunner, T., Zekveld, A., Sörqvist, P., Danielsson, H., Lyxell, B., … Rudner, M. (2013). The Ease of Language Understanding (ELU) model: theoretical, empirical, and clinical advances. *Frontiers in Systems Neuroscience, 7*(31), 1- 17. doi: 10.3389/fnsys.2013.00031

Rudner, M., Lunner, T., Behrens, T., Thorén, E. S. & Rönnberg, J. (2012). Working memory capacity may influence perceived effort during aided speech recognition in noise. *Journal of the American Academy of Audiology, 23*, 577-589. doi: 10.3766/jaaa.23.7.7

Saliasi, E., Geerlings, L., Lorist, M. M. & Maurits, N. M. (2013). The relationship between P3 amplitude and working memory performance differs in young and older adults. *PLos ONE, 8*(5), 1-9. doi: 10.1971/journal.pone.0063701

Sarampalis, A., Kalluri, S., Edwards, B. & Hafter, E. (2009). Objective measures of listening effort: Effects of background noise and noise reduction. *Journal of Speech, Language, and Hearing, 52*(5), 1230-1240. doi: 10.1044/1092-4388(2009/08-011)

Schrank, F. A. (2010). Woodcock-Johnson III tests of cognitive abilities. *Handbook of Pediatric Neuropsychology,* 1-20.

Shadle, C. H. (2007). Speech production and speech intelligibility. In M. Crocker (Ed.)'s Handbook of Noise and Vibration Control, John Wiley and Sons, Inc., Hoboken (pp. 293-300).

Shinn-Cunningham, B. G. & Best, V. (2008). Selective attention in normal and impaired hearing. *Trends in Amplification, 12*(4), 283-299. doi: 10.1177/1084713808325306

Shipstead, Z., Harrison, T. L. & Engle, R. W. (2015). Working memory capacity and the scope and control of attention. *Attention, Perception & Psychophysics, 77,* 1863-1880. doi: 10.3758/s13414-015-0899-0

Simon, S. S., Tusch, E. S., Holcomb, P. J. & Daffner, K. R. (2016). Increasing working memory load reduced processing of cross-modal task-irrelevant stimuli even after controlling for task difficulty and executive capacity. *Frontiers of Human Neuroscience, 10*(380), 1-13. doi: 10.3389/fnhum.2016.00380

Sörqvist, P. (2010). High working memory capacity attenuates the deviation effect but not the changing-state effect: Further support for the duplex-mechanism account of auditory distraction. *Memory & Cognition, 38*(5), 651-658. doi: 10.3758/MC.3835.651

Sörqvist, P. (2010). The role of working memory capacity in auditory distraction: A review. *Noise and Health, 12*(49), 217-224. doi: 10.4103/1463-1741.70500

Sörqvist, P. & Marsh, J. E. (2015). How concentration shields against distraction. *Current Directions in Psychological Science, 24*(4), 267-272. doi: 10.1177/0963721415577356

Sörqvist, P., Marsh, J. E. & Nöstl, A. (2013). High working memory capacity does not always attenuate distraction: Bayesian evidence in support of the null hypothesis. *Psychonomic Bulletin & Review, 20*(5), 897-904. doi: 10.3758/s13423-013-0419-y

Sörqvist, P., Nöstl, A. & Halin, N. (2012). Working memory capacity modulates habituation rate: evidence from a cross-modal auditory distraction paradigm. *Psychonomic Bulletin and Review, 19*(2), 245-250. doi: 10.3758/s13423-011-0203-9

Sörqvist, P., Stenfelt, S. & Rönnberg, J. (2012). Working memory capacity and visual-verbal cognitive load modulate auditory-sensory gaiting in the brainstem: Toward a unified view of attention. *Journal of Cognitive Neuroscience, 24*(11), 2147-2154. doi: 10.1162/jocn_a_00275

Sörqvist, P. & Rönnberg, J. (2014). Individual differences in distractibility: An update and a model. *Psych Journal, 3*(1), 42-57.

Spence, C. & Santangelo, V. (2010). Auditory Attention. In *Oxford Handbook of Auditory Science Hearing* (Vol. 3: Hearing, pp. 249-270). doi: 10.1093/oxfordhb/9780199233557.013.0011

Strauss, D. J. & Francis, A. L. (2017). Toward a taxonomic model of attention in effortful listening. *Cognitive, Affective, & Behavioral Neuroscience, 17*, 809-825.

Stroop, J. R. (1935). Studies of interference in serial verbal reactions. *Journal of Experimental Psychology, 18*(6), 643-662. doi: 10.1037/h0054651

Sur, S. & Sinha, V. K. (2009). Event-related potential: An overview. *Industrial Psychiatry Journal, 18*(1), 70-73.

Tan, J., Zhao, Y., Wang, L., Tian, X., Cui, Y., Yang, Q., Pan, W., Zhao, X. & Chen, A. (2015). The competitive influences of perceptual load and working memory guidance on selective attention. *PLoS ONE, 10*(6), 1-14. doi: 10.1370/journal.pone.0129533

Tiego, J., Testa, R., Bellgrove, M. A., Pantelis, C. & Whittle, S. (2018). A hierarchical model of inhibitory control. *Frontiers in Psychology, 9*(1339), 1-25. doi: 10.3389/fpsyg.2018.01339

Treisman, A. (1964). Selective attention in man. *British Medical Bulletin, 20*, 12-16. doi: 10.1093/oxfordjournals.bmb.a07274

Tremblay, P., Brisson, V. & Deschamps, I. (2021). Brain aging and speech perception: Effects of background noise and talker variability. *NeuroImage, 227*(117675), 1-16. doi: 10.1016/j.neuroimage.2020.117675

Trimmel, K., Schätzer, J. & Trimmel, M. (2014). Acoustic noise alters selective attention processes as indicated by direct current (DC) brain potential changes. *International Journal of Environmental Research and Public Health, 11*(10), 9938-9953. doi: 10.3390/ijerph111009938

Ulrich, R., Prislan, L. & Miller, J. (2021). A bimodal extension of the Erikson flanker task. *Attention, Perception & Psychophysics, 83*(2), 790-799. doi: 10.3758/s13414-020-02150-8

Unsworth, N. & Engle, R. W. (2007). The nature of individual differences in working memory capacity: Active maintenance in primary memory and controlled search from secondary memory. *Psychological Review, 114*(1), 104-132. doi: 10.1037/0033-295X.114.1.104

Vachon, F., Marsh, J. E. & Laboné, K. (2020). The automaticity of semantic processing revisited: Auditory distraction by a categorical deviation. *Journal of Experimental Psychology: General, 149*(7), 1360-1397. doi: 10.1037/xge0000714

Wilhelm, O., Hildebrandt, A. & Oberauer, K. (2013). What is working memory capacity, and how can we measure it? *Frontiers in Psychology, 24*, 1-22. doi: 10.3389/fpsyg.2013.00433

Wilson, R. H. (2003). Development of speech-in-multitalker-babble paradigm to assess word-recognition performance. *Journal of the American Academy of Audiology, 14*(9), 453-470.

Woldorff, J. M., Hackley, S. A. & Hillyard, S. A. (1991). The effects of channel-selective attention on the mismatch negativity wave elicited by deviant tones. *Psychophysiology, 28*, 30-42.

Woodcock, R. W., McGrew, K. S. & Mather, N. (2001). *Woodcock-Johnson III.* Rolling Meadows, IL: Riverside.

Zehr, J. & Schwarz, F. (2018). PennController for Internet Based Experiments (IBEX).

doi: 10.17605/OSF.IO/MD832